D1460502

Research Perspectives

Traffic loading on highway bridges

Peter Dawe

Published by Thomas Telford Publishing, Thomas Telford Ltd, 1 Heron Quay, London E14 4JD.
URL: www.thomastelford.com

Distributors for Thomas Telford books are
USA: ASCE Press, 1801 Alexander Bell Drive, Reston, VA 20191-4400, USA
Japan: Maruzen Co. Ltd, Book Department, 3–10 Nihonbashi 2-chome, Chuo-ku, Tokyo 103
Australia: DA Books and Journals, 648 Whitehorse Road, Mitcham 3132, Victoria

First published 2003

Front cover shows an abnormal load weighing approximately 860 tonnes crossing a concrete and beam slab bridge on the M53 near Ellesmere Port.
Photograph supplied courtesy of TRL Limited.

A catalogue record for this book is available from the British Library

ISBN: 0 7277 3241 2

Typeset by Academic + Technical, Bristol
Printed and bound in Great Britain by MPG Books, Bodmin, Cornwall

Contents

Author's biography xi
About TRL xiii
Foreword xv
Acknowledgements xvii
Conversion table xix

Chapter 1 **Introduction** 1
1.1 Historical overview 1
1.2 Scope of contents 2
1.3 Choice of loading model 3
1.4 Influences on loading rules 4
1.5 Development of standards 5
1.6 Future developments 7
References 10

Chapter 2 **Historical review: Part 1, 1911–54** **11**
2.1 Concrete Institute Report 1918 11
2.2 Ministry of Transport Standard Loading Train, 1922 13
2.3 Ministry of Transport Equivalent Loading Curve, 1931 13
2.4 BS 153—*Standard specification for girder bridges*, 1923 15
2.5 BS 153—*Standard specification for girder bridges*—1937 revision 17
2.6 *Code of practice for simply supported steel bridges*, 1949 18
2.7 Ministry of War Transport Memorandum No. 577, 1945 20
References 20

Chapter 3 **Historical review: Part 2, 1954–78** **21**
3.1 BS 153: Part 3A: 1954 21
3.1.1 Loading models 21
3.1.2 Normal loading curve 22

3.1.3 Short-span members 23
3.1.4 Concrete slabs 24
3.1.5 Notional traffic lanes 24
3.1.6 Abnormal loading 26
3.1.7 Braking forces 26
3.1.8 Footways 27
3.2 Ministry of Transport Memorandum No. 771, 1961 27
3.3 Ministry of Transport Interim Memorandum (bridges) IM 10, 1970 27
3.4 BS 153: Part 3A: 1972 28
3.5 Department of the Environment Technical Memorandum (bridges) BE 5/73 28
3.6 Interim design and workmanship rules—Part 1: Loading and general design requirements, 1973 30
3.7 Department of Transport Technical Memorandum (bridges) BE 1/77, 1977 30
References 30

Chapter 4 **Limit state design** **33**
4.1 BS 5400: Part 2: 1978 33
4.2 Limit state design 33
4.3 HA loading curve 34
4.4 HB vehicle 35
4.5 Application of HA and HB loading 36
4.6 Partial load factors 36
4.7 Load combinations 37
4.8 Centrifugal force 37
4.9 Braking loads 38
4.10 Skidding loads 40
4.11 Collision loads 40
4.12 Department of Transport, *Departmental standard BD 14/82*, 1982 41
4.13 Department of Transport's *Interim revised loading specification*, 1982 42
References 43

Chapter 5 **Revision of HA Type loading—1. Short span** **45**
5.1 Background 45
5.2 Methodology 46
5.2.1 C&U vehicle loading and impact 46
5.2.2 Overloading 47
5.2.3 Lateral bunching 48
5.3 Derivation of loading 50
5.4 Future European vehicles 51
References 52

Chapter 6 **Revision of Type HA loading—2. Long span** **55**
6.1 Background 55
6.2 Outline of loading study 56

6.3 Data used in study 57
6.3.1 Vehicle weights and lengths 57
6.3.2 Proportions of different vehicle types 58
6.3.3 Variation in average vehicle weights through the day 58
6.3.4 Spacing between vehicles 59
6.4 Characteristics of traffic jams 59
6.5 Vehicle lane selection 60
6.6 Projections of traffic growth to 1990 60
6.7 Generation of traffic streams 61
6.8 Determination of equivalent uniformly distributed load 62
6.8.1 Method 62
6.8.2 Normal distribution 63
6.8.3 Extremal distribution 63
6.8.4 Lane load factors 64
6.9 Sensitivity studies 65
6.9.1 Percentages of heavy goods vehicles 65
6.9.2 Jam frequencies and duration 66
6.9.3 Traffic flow rate 66
6.9.4 Vehicle spacing 66
6.9.5 Lane distribution 66
6.10 Design values 66
References 68

Chapter 7 **Composite version of BS 5400: Part 2** **69**
7.1 Background 69
7.2 Type HA loading 69
7.3 Type HB loading 70
7.4 HA lane factors 70
7.5 Load combinations and partial factors 73
7.6 Accidental wheel loading 73
7.7 Loads due to vehicle collisions with parapets 73
7.8 Secondary live loads 73
7.9 Department of Transport implementation 74
7.10 *Design manual for roads and bridges*, BD 37/01 74
References 74

Chapter 8 **Short-span assessment loading** **75**
8.1 Background 75
8.2 Outline of methodology 76
8.3 Static load models 77
8.3.1 Vehicle data 77
8.3.2 Traffic modelling 77
8.4 Dynamic load model 78
8.4.1 Data 78
8.4.2 Dynamic effect model 78
8.5 Probabilistic load model 79

8.6 Reliability analysis 79
8.7 Lateral bunching 80
8.8 Load reduction factors 80
References 82

Chapter 9 **Loading for assessment** **83**
9.1 Introduction 83
9.2 Technical Memorandum (bridges), BE 4 84
9.2.1 Background 84
9.2.2 Vehicle trains 84
9.2.3 Application of vehicle trains 85
9.2.4 Public service vehicles 86
9.2.5 Load distribution 86
9.2.6 Operation bridgeguard 86
9.3 Technical Memorandum (bridges) BE 3/73 86
9.4 Departmental standard BD 21/84 87
9.4.1 Background 87
9.4.2 Reduced loadings 87
9.4.3 Single axle and single wheel loads 89
9.4.4 Application of loading 89
9.4.5 Centrifugal effect 89
9.5 Advice note BA 16/84 90
9.6 Amendment No. 1 to BD 21/84 90
9.6.1 Background 90
9.6.2 Assessment loading 91
9.6.3 Wheel loading 91
9.6.4 Special structures 92
9.6.5 Lane factors 93
9.7 *Design manual for roads and bridges* BD 21/93 93
9.8 *Design manual for roads and bridges* BD 21/97 93
9.8.1 Adjustment factors 94
9.8.2 Special structures 95
9.8.3 Lane factors 95
9.9 *Design manual for roads and bridges* BA 16/97 95
9.10 *Design manual for roads and bridges* BD 21/01 96
References 96

Chapter 10 **Abnormal loading** **99**
10.1 Introduction 99
10.2 Regulation of abnormal vehicles 100
10.3 STGO vehicles 100
10.4 SO vehicles 101
10.5 Review of studies of abnormal vehicle loading 101
10.5.1 *Highway bridge loading appraisal* 101
10.5.2 *Long span bridge loading*: Contractor Report 16 102
10.5.3 *Revalidation of traffic loading for abnormal vehicles* 104

10.6 Loading models for STGO vehicles 105
10.7 Loading model for SO vehicles 106
10.8 Development of loading model for abnormal vehicle effects 108
10.8.1 Background 108
10.8.2 Shortcomings of existing HB model 108
10.8.3 Development of SV loading model 109
10.8.4 Overload factor (OF) 110
10.8.5 Dynamic amplification factor (DAF) 111
10.8.6 Partial load factors 113
10.8.7 Masonry arches 113
10.9 *Design manual for roads and bridges* BD 86/01 113
10.9.1 Background 113
10.9.2 Loading 114
10.9.3 Application 114
10.9.4 Special structures 114
10.9.5 Reserve factors 114
10.9.6 HB-to-SV conversion charts 115
10.9.7 Management of STGO vehicle movements 115
References 115

Chapter 11 **Collision loads** **117**
11.1 Introduction 117
11.2 Early requirements 117
11.3 Bridge rehabilitation programme 118
11.4 Study of heavy goods vehicle collisions with highway structures 119
11.5 *Design manual for roads and bridges* BD 48/93 120
11.6 *Design manual for roads and bridges* BD 60/94 121
11.7 Computer simulation of heavy goods vehicles collisions with bridges 121
References 123

Chapter 12 **Control of traffic loading** **125**
12.1 Introduction 125
12.2 Vehicle regulations 125
12.3 European vehicles 128
12.4 Traffic loading restrictions 130
12.5 Weight restrictions 130
References 131

Chapter 13 ***Eurocodes*** **133**
13.1 Background 133
13.2 Development of ENV 1991-3 134
13.3 ENV 1991-3 [2], scope and contents 134
13.3.1 Load Model 1 134
13.3.2 Load Model 2 134
13.3.3 Load Model 3 135

13.3.4 Load Model 4 137
13.4 Derivation of Load Models 1 and 2 137
13.4.1 Background 137
13.4.2 Methodology 137
13.4.3 Extrapolation methods 138
13.4.4 Dynamic effects 139
13.4.5 Loading patterns 140
13.4.6 Influence lines 141
13.4.7 Calibration of Load Model 1 142
13.4.8 Load Model 2 142
13.5 Application of loading 142
13.5.1 Lane widths 142
13.5.2 Partial live load factors 143
13.5.3 Load combinations 143
13.5.4 Reduction factors 143
13.6 Other types of loading 143
13.6.1 Accidental loads 143
13.6.2 Braking 144
13.6.3 Centrifugal force 144
References 145

Chapter 14 **National Application Document (NAD)** **147**
14.1 Background 147
14.1.1 National requirements 147
14.1.2 Calibration studies 147
14.2 Initial parametric calibrations 148
14.2.1 Scope 148
14.2.2 Congested traffic 148
14.2.3 Flowing traffic 148
14.2.4 Adjustment factors—Load Model 1 149
14.2.5 Adjustment factors—Load Model 2 150
14.2.6 Reduction factors 150
14.2.7 Abnormal vehicles—Load Model 3 151
14.3 Final parametric calibrations 152
14.3.1 Scope 152
14.3.2 Adjustment factors (α) 152
14.3.3 Reduction factors (ψ) 153
14.3.4 Abnormal vehicles—Load Model 3 153
14.3.5 Secondary loads—braking 154
14.3.6 Secondary loads—centrifugal forces 154
14.3.7 Accidental loads—collision forces on piers 154
14.3.8 Accidental loads—vehicles on footways 154
14.4 Published version of the UK NAD 155
14.4.1 Introduction 155
14.4.2 Partial safety factors (γ) 155
14.4.3 Adjustment factors (α) 155

14.4.4 Reduction factors (ψ) 156
14.4.5 Tables of factors 157
References 157

Appendix **Use of OPTAX program** **159**
A1 Introduction 159
A2 Vehicle trains 159
A3 Vehicle spacing 160
A4 Impact factors 160
Reference 160

Index 161

Author's biography

Peter H. Dawe, CBE, MA, MSc, CEng, MICE

Educated at University College School, Hampstead, and, after National Service in the RAF, read Mechanical Sciences at Emmanuel College, Cambridge, from 1954 to 1957. On graduation worked for a short time with the De Havilland Aircraft Company before joining the Chief Civil Engineer's Department of the Eastern Region of British Rail where, after a two-year graduateship, joined the Bridge Office at King's Cross. In 1963 left British Rail to take an MSc at the Institute of Sound and Vibration Research, Southampton University. On completion joined the contractor Taylor Woodrow, spending a year in the Structures Research Laboratory before working on site on the reconstruction of Euston Station.

Joined the, then, Ministry of Transport in 1967, spending the first nine years in the Highways Engineering Computer Branch involved in the development and evaluation of structural computer programs. In 1976 transferred to the Bridges Engineering Division, becoming head of the Division in 1986. From then, until retirement in 1994, was responsible for the engineering standards and policy for the design, construction, maintenance and assessment of all types of structure on the trunk road network. Much of the work was directed towards the preparation of the technical and procedural documentation necessary to support the Department of Transport's 15-year bridge rehabilitation programme.

Creating the future of transport

TRL Limited is one of the largest and most comprehensive independent centres for the study of transport in the world. For more than 70 years it has been at the forefront of providing research?based technical help to the highest standards of quality, reliability and impartiality, to enable clients to solve of a range of complex problems that arise within the transport arena.

TRL's work continues to enable government customers to set standards for highway and vehicle design, formulate policies on road safety, transport and the environment, and encourage good traffic engineering practice. The company also provides services to other clients in the UK and overseas, providing research and implementation advice. These customers include local and regional authorities, major civil engineering contractors, transport operators, foreign governments and international aid agencies.

Foreword

There are over a hundred thousand highway bridges in the UK, many of which were constructed before the advent of the motor vehicle. Until the middle of nineteenth century the loads carried by these structures were only a small proportion of their dead load. It was not until the development of the steam traction engine that engineers were forced to consider traffic loads, although the first national standard for traffic loads on highway bridges was not published until 1918.

Peter Dawe, a former Head of the Bridges Engineering Division of the Department for Transport, has done an excellent job of charting the evolution of traffic loading rules from their introduction up to the present day, including the development of Eurocodes, which will eventually supersede national standards. There have been many changes in the intervening years, which have affected the way traffic loads are modelled. In particular there have been increases in the volume of traffic and the weight of individual vehicles, changes in the mix of traffic, and increases in the width and length of bridge spans. This has led to the introduction of statistical methods to accurately model traffic characteristics. In addition to the development of rules for normal traffic, it has also been necessary to develop models that represent the effects of abnormal loads.

There are many aspects of loading to be considered. As well as static gravitational loads (including overloaded vehicles) engineers also have to consider dynamic loads, braking forces, and traction and centrifugal forces on structures that are curved in plan. The book examines each of these loads in some detail.

In recent years there has been considerable emphasis on the load carrying capacity of the existing bridges, particularly those that were designed to codes that have subsequently been superseded. This has required engineers to consider whether rules derived for the design of new structures are appropriate for assessing existing structures. As a result the rules have been refined to allow the engineer to take account of the site-specific loading where appropriate.

The book gives a detailed description of how all the factors that affect traffic loads have evolved and been drawn together to produce the rules in use today. This will be of value to students who wish to delve in to the background behind the current loading rules, practicing engineers who wish to understand the premises upon which the rules are based and researchers who wish to develop the subject further.

The Department of Transport and, latterly, the Highways Agency, have sponsored an extensive programme of R & D and consultancy work over a number of years to develop the UK design and assessment traffic loading models. TRL has had a significant role in this work and I am grateful to them for taking the initiative in producing a book which I am sure will be an important reference source for the engineering community.

Ginny Clarke
Chief Highways Engineer
Highways Agency

Acknowledgements

The authors wish to acknowledge the following organisations and companies for permission to reproduce the following figures and tables.

British Standards Institute
Figure 2.4 (BSI 153, 1925)
Figure 2.5 (BSI 153, 1937)
Figure 3.2 (BSI 153, 1954 part 3A)
Figure 3.3 (BSI 153, 1954 part 3A)
Figure 3.4 (BSI 153, 1972 part 3A)
Figure 4.1 (BSI 5400, part 2 1978)
Figure 13.1 (BSI DD ENV 2000 1991-3)
Figure 13.2 (BSI DD ENV 2000 1991-3)
Figure 13.3 (BSI DD ENV 2000 1991-3)
Table 4.1 (BSI 5400, part 2 1978)
Table 4.2 (BSI 5400, part 2 1978)
Table 13.1 (BSI DD ENV 2000 1991-3)
Table 13.2 (BSI DD ENV 2000 1991-3)
Table 13.3 (BSI DD ENV 2000 1991-3)

Concrete Institute
Figure 2.1

Department for Transport
Figure 2.3
Figure 4.3
Figure 5.1
Figure 5.2
Figure 7.1
Figure 8.2

Figure 9.1
Figure 9.2
Figure 9.3
Figure 9.4
Figure 9.5
Figure A1
Table 5.1
Table 5.2
Table 5.3
Table 5.4
Table 7.1
Table 7.2
Table 9.1
Table 12.2

Highways Agency
Table 1.1

Her Majesty's Stationery Office
Figure 2.2
Figure 9.6
Table 12.1
Table 12.3

Institution of Structural Engineers
Figure 2.6

Finally, TRL Limited gratefully acknowledges the contribution made by the Highways Agency in the writing of this book.

Imperial/metric conversion chart

Recommended list of units, unit abbreviations, quantity symbols and conversion factors.
SI base units, derived units and multiples.

Quantity and symbol	Units and multiples	Unit abbreviations	Conversion factors for existing units	Remarks
Length (various)	kilometre	km	1 mile = 1.609 km	
	metre	m	1 yard = 0.9144 m	
	millimetre	mm	1 ft = 0.3048 m	
	micrometre	μm	1 in = 25.40 mm	1 micrometre = 1 micron
Area (A)	square kilometre	km^2	1 $mile^2$ = 2.590 km^2	
	square metre	m^2	1 yd^2 = 0.8361 m^2	
	square millimetre	mm^2	1 ft^2 = 0.09290 m^2	
			1 in^2 = 645.2 mm^2	
Volume (V)	cubic metre	m^3	1 yd^3 = 0.7646 m^3	
	cubic centimetre	cm^3	1 ft^3 = 0.02832 m^3	To be used for solids and liquids
	cubic millimetre	mm^3	1 in^3 = 16.39 cm^3	
			1 UK gallon = 4546 cm^3	
Mass (m)	megagram	Mg	1 ton = 1.016 Mg	Megagram is the SI term
	(or tonne)	(t)		
	kilogram	kg	1 lb = 0.4536 kg	
	gram	g		
Unit weight (γ)	kilonewton per cubic metre	kN/m^3	100 lb/ft^3 = 15.708 kN/m^3 (62.43 lb/ft^3 pure water = 9.807 kN/m^3 = specific gravity 1.0 approx.)	Unit weight is weight per unit volume
Force (various)	meganewton	MN	1 tonf = 9.964 kN	
	kilonewton	kN	1 lbf = 4.448 N	
	newton	N	1 kgf = 9.807 N	
Pressure (p, u)	meganewton per square metre	MN/m^2	1 $tonf/in^2$ = 15.44 MN/m^2	To be used for shear strength, compressive strength, bearing capacity, elastic moduli and laboratory pressures of rock
	megapascal	MPa	(1 MN/m^2 = 1 N/mm^2)	
Stress (σ, τ) and Elastic moduli (E, G, K)	kilonewton per square metre	kN/m^2	1 lbf/in^2 = 6.895 N/m^2	Ditto for soils
			1 lbf/ft^2 = 0.04788 kPa	
	kilopascal	kPa	1 $tonf/ft^2$ = 107.3 kPa	
			1 bar = 100 kPa	
			1 kgf/cm^2 = 98.07 kPa	

Chapter 1 Introduction

1.1 Historical overview

Despite the fact that bridges of one form or another have been constructed for thousands of years, it is only comparatively recently that the bridge builder has been concerned about quantifying the loading to be carried by the bridge. This may seem a little surprising considering that the reason for any bridge to be built is that it should carry some form of live load. However, in the earliest times the traffic was light and would have been insignificant in comparison with the dead load of the structure itself. Moreover, even if the builders had thought about the live loading, it is doubtful whether their understanding of structural behaviour would have been sufficient for them to do anything about the problem.

It was the development of the steam traction engine in the latter half of the nineteenth century that led to the development of live loading requirements for highway bridges. When these vehicles first appeared they would be travelling on roads and bridges which would probably have carried nothing heavier than a horse and cart. It was soon recognised that with the advent of heavier vehicles there was a need for some method of determining the carrying capacity of highway bridges, although some bridge owners responded by restricting the use of their bridges to the traffic of the district. Some of the earliest Acts of Parliament which regulated the weights of vehicles were more concerned with preventing damage to the road surface than with making sure that bridges weren't overloaded. Although petrol driven motor vehicles were making their appearance at the turn of the century, these were mainly cars being used for pleasure and most of the bridges were more than adequate to carry them. However, the use of heavier vehicles was increasing and, as will be seen in Chapter 2, some local authorities had actually been developing their own requirements for the design loads for bridges in their localities.

The 1914–18 war saw a large increase in the use of motor vehicles, especially for the transport of goods and equipment. Motor vehicles were being used to carry goods for longer distances, rather than just as a means of transporting goods for the short journey to and from the local railway goods yard. Thus the need for a national approach to the problem of access to bridges was becoming more urgent in order to prevent any restrictions on traffic movements by local bridge owners imposing their own rather arbitrary weight restrictions. This national approach was brought about through the creation in 1919 of the Ministry of Transport who, a few years later, introduced their Standard Loading Train, as described in Chapter 2. Although this was based on a typical steam powered tractor pulling three trailers, it would have been more than adequate to cover the motor vehicles of the day. This was the first national traffic loading requirement in the UK; details of this loading and subsequent developments for traffic up to the present day are described in the later chapters. Table 1.1 at the end of this chapter gives an outline of the history of the development of the various traffic loading rules.

1.2 Scope of contents

This review of loading rules in the UK is purely concerned with the loads which arise as a result of the vehicular traffic being carried by highway bridges. A major part is concerned with the gravitational loads from normal traffic, but the loading from abnormal or special vehicles is also covered. Other forms of loading attributable to motor vehicles such as the horizontal forces from braking, collisions with structures and centrifugal effects are also covered. Although the basic loading requirements are important, the review also covers the equally important aspect of their method of application. Vehicle loading is often required to be applied in combination with other forms of loading, such as those due to wind or differential temperature, and reference is made to these combinations although the requirements for these other forms of loading are not covered.

In reading the review it will be noted that quite often a particular code or standard will be revised or amended. In these cases only the changes from the original have been described in the text; it can be assumed that other requirements remain as they were in the original or previous version. References to the documents which have been used in the review are given at the end of each chapter, and even if a document is referred to in more than one chapter it is still listed at the end of each chapter. There are two unpublished papers which have been particularly helpful in tracing the development of the various codes and providing background material for several of the chapters. These are the reports by Henderson [1, 2] for the Department of Transport written at the time of the development of the loading requirements for BS 5400 [3], and the report by McMahon and Brady [4] of TRL Limited reviewing the design loads for buried structures. It should be noted that a number of the other documents referred to are so far unpublished and this is noted in the reference.

In addition to looking at the live loading to be used for the design of new structures, the review also looks at the loading criteria for assessing the load carrying capacity of existing structures. Assessment loadings, as a part of a comprehensive assessment code, were first introduced about ten years after the end of the 1939–45 war when it became evident that there were a considerable number of weak bridges in the UK. Many of these would have been designed to much earlier standards and would have suffered from heavy usage and very little maintenance during the war years. The assessment loading is, in some ways, more complex than loading for design in that it not only has to help determine whether a structure can safely carry the traffic of the day, but it also has to determine the levels of restricted loading which a sub-standard structure can safely carry. It will be seen that the more recent versions of the assessment codes recognise the benefit of taking account of site specific and structure specific factors when dealing with an existing structure. As well as being able to take account of the actual state and condition of the structure in question there are also advantages in being able to take account of the actual make up and volume of traffic using the particular structure. This is slightly different from the approach in design where the requirements tend to be on the conservative side in order to cover a range of unknown situations.

The review shows how it has been recognised that there is a strong link between the loading models, which represent the effects of the current and future traffic, and the rules which are used to control the construction of vehicles and their use on the public highway. Highway bridges are long-life structures which are designed to have adequate and acceptable margins of safety, but this situation can only be maintained if there are matching controls on the weights and dimensions of vehicles allowed to use them. *The Construction and Use (C&U) Regulations* used for this purpose are described in Chapter 12, together with the method used for determining the acceptability of new vehicle designs. The regulations governing the use of abnormal vehicles are described in Chapter 10.

Although there are figures in the text which allow a comparison to be made between various different loading criteria, no attempt has been made to assess the effect of changes in loading requirements on the design of structures themselves. It should be remembered that the live traffic loading is only one of the factors which influences the design of the structure and that other forms of loading, such as dead load, wind and temperature, can be equally or more significant. When comparing the effects of different design codes it is important that the comparisons should be done using the complete design package and comparing the finished designs. This is particularly important when comparing codes, and their loading requirements, which have different approaches to design, such as permissible stress and limit state. Here the use of an implied global factor of safety on the one hand and the use of partial factors on the other make a straight comparison of the loading criteria almost meaningless.

1.3 Choice of loading model

It is generally accepted that the representation of actual traffic for use in design must be on the basis of a model of one sort or another. Although the first national load model was based on a specific train of vehicles, it became obvious that this required a lot of computational effort for its application and that it was fairly inflexible with its defined axle spacings. The Standard Loading Train was soon to be replaced by the concept of an equivalent uniformly distributed load and an associated knife-edge load (KEL). This combination was formulated in such a way that for varying spans it would apply the same total loading on the structure as the relevant axles of the train. Later on it was realised that a more accurate way of modelling traffic was to model the effects of the loading in terms of bending moments, shears and reactions, albeit retaining the form of the model as a uniformly distributed load plus a KEL. This concept has continued to be used up to the present day although, as described in Chapter 13, the draft Eurocode [5] adopts a tandem axle arrangement in place of the single KEL. However, whereas the earlier models were based on idealised arrangements of vehicles operating at their legal weights and with rather crude allowances for dynamic effects, the current loading models are based on the actual weights of vehicles in service, including any overloading, and a more realistic evaluation of possible dynamic effects.

One advantage of the distributed plus KEL approach is that it is comparatively easy to apply to a structure and considerably reduces the amount of computation. It also means that the model is fairly flexible in that it can cope with a whole range of different vehicles and combinations of vehicles. However, there are difficulties in using a constant single axle load in association with a distributed load which varies with the span. It turns out that slightly different models are required to model both bending moments and shear forces with the same degree of accuracy; one way to overcome this would be by having different values of the KEL for moments and shears respectively. Moreover the use of a distributed load is not the most effective way of modelling the effects of wheel loads on short spans. In the first case it has been accepted that it is better to stick with a single KEL and a single curve for the distributed load, and to accept that there will be some loss of accuracy in basing the load model on the worst of the moment and shear effects. In the second case it has been found necessary to introduce a single heavy wheel or axle for loading small elements of a bridge deck or short span members.

It will be noted, when following the development of the various load models, that although the so called 'formula loading' is relatively easy to apply, a considerable amount of work is necessary to develop the loading models themselves to ensure that they are truly representative of the effects of actual traffic and that they cover the various traffic scenarios. Also, besides the basic load models it is also necessary to develop rules for the application of the loading which take account of driver behaviour, for instance lane selection and the effect of temporary contra-flows. Thus although the design effort may be reduced as much as possible, a lot of work is required from the code writers and researchers to bring this about.

It is interesting to note that although the original loading train model, which was in fact the abnormal vehicle of the day, is no longer used to model the effects of normal vehicles, a single vehicle model is still used to model the effects of abnormal vehicles. This vehicle with certain modifications has remained almost unchanged since it was first introduced in the UK, despite increases in the number and configurations of such vehicles on the road network. The *Eurocode* [5], on the other hand, lists 17 special vehicles (SVs) which may be used to model the different types of abnormal vehicles currently in use on the roads of Europe.

1.4 Influences on loading rules

In following the development of traffic loading models and the associated application rules it will be noticed that a number of factors have influenced the development and presentation of such rules. Strangely, though, the major change in design philosophy from permissible stress to limit state design in the late 1970s had very little impact at the time on either the form of the basic load model or the values of the loading. As will be discussed later, increases in allowable vehicle and axle weights have not been very significant in causing changes in the values of the basic load models. As far as traffic is concerned, changes and enhancements of the load models have been as a result of the large increases in the amount of traffic and in particular the number of heavy goods vehicles now in use. However, increases in vehicle weights have brought about changes in the secondary horizontal loads, such as braking and centrifugal effects, which are directly related to the weight of a vehicle.

A major influence on presentation and coverage of the loading standards has been the emergence, largely as a result of computers, of much more powerful methods of structural analysis. The earlier codes went into great detail to describe how loads should be distributed in the structural forms in use at the time, for example by including empirical rules about the loads to be applied to the edges of slabs and to cross-members. These rules are no longer necessary as modern computerised methods of structural analysis can easily distribute whatever loads are applied to the structure in question. However, although the process of load distribution has become more accurate and less laborious, the application of the loading has become more complex and difficult. In the early days most bridges would only be wide enough to carry two lanes of traffic, one in each direction. Nowadays a bridge on a major road could be carrying four or even more lanes on each carriageway. In addition, it is now usual practice to adopt contra-flow working when repairs and maintenance are being carried out. Thus the designer now needs guidance on how the loading is to be applied to each lane, and must also consider a much larger number of loaded lane configurations when carrying out the analysis of the structure.

The advent of computers and electronic devices has provided not only the means to capture much more data about the make-up of traffic and information on individual vehicles, but has also provided the means to process such information. For instance, weigh-in-motion devices linked to computers can record and store information about very large numbers of actual vehicles in comparatively short periods. Thus the development of loading models can be based on real vehicles rather than idealised ones, and the amount of data allows a statistical approach to the way that the data are handled. The deterministic approach adopted for the earlier codes, which included a fair amount of engineering judgement to deal with various unknowns, has given way to a more scientific and research backed approach. However, even in the development of the latest codes there are still gaps in the knowledge about traffic behaviour, although the use of computers now makes it fairly easy to carry out sensitivity studies to assess the significance of the various assumptions that have been made.

1.5 Development of standards

The development of traffic loading models and rules has involved central government, local authorities and the engineering professions. The first efforts were very much local matters with individual local authorities adopting loading requirements for their bridges which were appropriate for their locality and the local needs. Thus, the first attempt at the development of a national set of rules involved mainly municipal engineers as well as an industrial organisation [6]. However, it wasn't long before central government, in the form of the newly created Ministry of Transport, actually introduced the first national loading model [7]. About the same time the British Engineering Standards Association (later to become the British Standards Institution) became involved in the preparation of a standard for the design of steel bridges which included highway loading [8]. The committee responsible for the drafting included representatives from government departments, professional institutions and scientific and industrial organisations.

Thereafter, both the Ministry of Transport (MoT) and the British Standards Institution (BSI) were involved in producing standards for bridge design, including the loading requirements. It should be remembered that the Ministry of Transport had a responsibility for the main routes in the UK, whereas the British Standards Institution had a wider interest in providing for the design and construction of bridges overseas as well. The major professional institutions became more fully involved when a joint committee of the Institutions of Civil and of Structural Engineers produced a code of practice for steel bridges [9] which made reference to both the MoT and BSI loading rules. This duplication of standards was to continue for some time until the publication of BS 153: Part 3A [10] in 1954. This was formally adopted by the MoT for use on its own roads, and although the Ministry had been fully involved in the preparation of BS 153 it still had some additional requirements for its own schemes.

This procedure by the Ministry (later to become the Department) of Transport for formally adopting British Standards for bridge design, but with its own additional amendments or requirements, has continued to the present day. The transition from permissible stress design to limit state design, through the use of the ten-part comprehensive bridge code BS 5400 for the design of steel, concrete and composite bridges, was fully accepted and implemented by the Department, though here again they still had their own additional requirements. In the case of traffic loading, BS 5400: Part 2 [11] was only implemented for spans up to 40 m pending the completion of a review by the Department of long-span loading requirements. This review, and a concurrent review of short-span loading for a new assessment code, led to a complete revision of the loading requirements for normal vehicles. The revision proposed by the Department was discussed and agreed with the British Standards Institution through one of their committees which contained representatives of a number of organisations with an interest in bridge design and construction. However, as will be described in Chapter 7, the amendments were actually printed and issued by the Department acting on behalf of the BSI. Thus, after many years, a single agreed set of requirements for traffic loading on bridges was available for all bridges designed in the UK.

It is perhaps not surprising that the DoT should wish to have such a major say in the question of the traffic loading on highway bridges. Although the Department, and now the Highways Agency, had responsibility for less than ten per cent of the road network in the UK, their motorways and all-purpose trunk roads carried over a third of all traffic and over half of all heavy goods traffic. Thus the Department had a vested interest in ensuring that the loading to which its bridges were designed was fully adequate for the traffic they had to carry, and that these major routes were not restricted unnecessarily by the need for unforeseen repairs or strengthening works. However, the interest and involvement of the Department in the production and contents of standards was also partly due to the economic circumstances at the time. With an increase in competition, private and professional organisations could no longer afford to devote the time and resources to the development of codes and standards which, up to then, they had done on a mainly voluntary basis. However, they still continued to contribute to the review and finalising of such documents. It was thus left to the Department to organise and fund much of the research and development work that was necessary to keep standards up to date and as relevant as possible. This state of affairs has continued with the UK contribution to the development of the *Eurocodes*, including the one which deals with bridge loading, where the Highways Agency has funded consultants to undertake background studies and to participate in a number of the project groups.

In this review the Department of Transport (DTp) has been referred to by its name at the time of whatever is being discussed. Thus in the early days the central government body is referred to as the Ministry of Transport (MoT), though at one time later transport matters came under the auspices of the Department of the Environment (DoE). Since 1994 the responsibility for highway and bridge standards has been transferred to the Highways Agency.

1.6 Future developments

The next stage regarding the traffic loading to be used for design in the UK will be the adoption of the European standard as described in Chapter 14. Although the loading code, and the other codes which form an integral part of the complete structural design package, will ensure that there is a common approach to structural design throughout Europe, there will still be scope for each member state to tailor the codes to meet their own particular requirements. The main traffic loading model in the *Eurocode* is one which is familiar to UK designers, as it involves both a distributed load and concentrated loads. It will be interesting to see how long this approach remains the norm or whether there might be a switch back to the original idea of a train of specific vehicles. The power and speed of computers is increasing at such a rate that it may soon be possible and feasible to generate site specific loading for each structure using basic traffic data collected either at the site or in the near vicinity.

Highway bridges are long-life structures so there is always the question when developing a traffic loading model as to what allowance to build into it to allow for unforeseen changes in the traffic. While there is usually some form of control over the weight and dimensions of individual vehicles, it is much more difficult to control the volume and the make-up of the traffic of which they are a part. The latter could be significant for the longer span structures where the percentage of heavy goods vehicles crossing a particular structure could exceed the value assumed

Table 1.1. Development of traffic loading rules

Date	Event/publication	Comment
End of 19th century		Principal live loading on bridges deemed to be due to crowd loading. UDL used for design of bridge decks, for example 100 lb/ft^2 (4.8 kN/m^2) for Hungerford Suspension Bridge.
1904	Restriction on vehicle weights	8 ton limit for single axle, 12 ton limit for gross vehicle weight.
1911	Joint committee set up to examine loads on highway bridges (report issued by Concrete Institute, 1918)	Recommended three loading classes. Wheel loads assumed to disperse through concrete at 45°.
1918	Ministry of Transport (MoT) established	
1922	MoT *Report on administration of the Road Fund for year 1921–1922*	First UK standard for loads on highway bridges: Standard Loading Train (SLT) included 50 per cent allowance for impact. Loads derived deterministically. Need recognised for higher loadings in certain localities to be determined by Local Authority.
1923	BS 153 *Part 3: Loads and stresses*	Traffic live loading to be specified by the Engineer. Impact factor inversely proportional to span.
1925	BS 153 Appendix 1 to Part 3, BSI	Unit loading train (similar to SLT) for highway girder bridges. Recommended 15 units for all road bridges in Great Britain.
1931	MoT *Standard loading for highway bridges*	Standard Loading Curve. Formulaic (deterministic) approach using equivalent UDL and KEL, with allowance for impact. Heavy wheel load introduced for short-span structures.
1937	BS 153 Part 3 (1st revision)	Introduced Types A and B loading. Impact allowance varied with span.
1945	MoWT *Memorandum on bridge design and construction*, Memo No. 577	Loading identical with 1931 SLC requirements.
1949	Joint committee of ICE and IStruct.E, *Code of practice for simply supported steel bridges*	Appendix B covers BS unit loadings (Types A and B), MoT equivalent loading, and proposes abnormal loading (Type C). Loading varied with number of traffic lanes. Impact allowance varied with span. Consideration given to longitudinal forces generated by braking vehicles.
1954	BS 153: Part 3A (2nd revision)	Appendix A introduces Types HA and HB loading. HA comprises deterministic formula loading based on 22-ton vehicles, and an alternative wheel loading. HB loading with axle number and spacing based on typical abnormal trailers of the day; axle loads are heaviest allowed by law.
1961	MoT *Standard highway loadings*, Memo No. 771	Formal adoption of BS 153: 1954 by MoT but with some additional requirements. Applicable to all types of highway bridge.
1966	BS 153: Part 3A (reissue)	As Memorandum No. 771.

Table 1.1. Continued

Date	Event/publication	Comment
1970	MoT interim memorandum (bridges) IM 10, *Standard highway loadings*	Supplement to Memorandum No. 771. Guidance given on the design of substructures.
1972	BS 153: Part 3A (metricated)	Metrication resulted in some slight changes to the configuration and axle weight of the HB abnormal model.
1973	DoE technical memorandum (bridges) BE 5/73, *Standard highway loadings*	Loads applicable to all highway structures except steel box girders. Required a minimum of 30 units of HB loading for public roads. HA UDL capped at 31.5 kN for loaded lengths up to 6.5 m. HA wheel load and HB loading assumed to cover design of short spans.
1973	Interim Design and Workmanship Rules. *Part 1: Loading and general design*	Loading requirements for steel box girders. Limit state design. Nominal loads based on BS 153: Part 3A: 1972.
1977	DTp Technical memorandum (bridges) BE 1/77, *Standard highway loadings*	Revision of BE 5/73. Highway structures defined as having a span or internal diameter greater than 0.9 m.
1978	BS 5400 Part 2, *Specification for loads*	Introduction of limit state design. HA loading based on 24-tonne vehicles. HA UDL capped at 30 kN/m for loaded lengths up to 30 m. Minimum UDL intensity now required to be 9 kN/m. Minimum of 25 units of HB loading required for public roads. HB loading (and HA wheel load) assumed to cover design of short spans.
1982	DTp BD 14, *Loads for highway bridges*	Implemented BS 5400: Part 2 for loaded lengths up to 40 m.
1982	DTp Interim revised loading specification	Revised and much enhanced HA UDL for spans from 40 m to 400 m.
1984	DTp BD 21, *The assessment of highway bridges and structures*	HA loading re-derived for Construction and Use vehicles, taking into account effects of overloading, lateral bunching and impact factor of 1.8. Loading derived for full range of spans (i.e. no longer capped for short spans).
1988	DTp BD 37, Loads for highway bridges (composite version of BS 5400: Part 2). Incorporated in DMRB in 2001.	Revision of BS 5400: Part 2: 1972 containing revised HA loading; short span based on BD 21/84, enhanced long span derived statistically from live traffic data. Covers spans up to 1600 m.
1994	CEN, ENV 1991–3. *Eurocode 1: Basis of design and actions on structures. Part 3: Traffic loads on bridges*	European pre-standard for traffic loads on bridges. Covers spans up to 200 m. Constant UDL for all spans and tandem axle systems. 3 m notional lanes.
1997	HA, BD 21. The assessment of highway bridges and structures. Revised in 2001	Revision of short span assessment loading by statistical methods and allowing for site factors for volumes of traffic and road surface condition.
2000	BSI, DD ENV 1991–3: 2000	Issue of ENV 1991–3 together with UK NAD. Constant UDL for all lanes across carriageway.

Table 1.1. Continued

Date	Event/publication	Comment
2001	HA, BD 86. The assessment of highway bridges and structures for the effects of STGO and SO vehicles	Load models for assessing the effects of STGO and SO vehicles based on real vehicles. Less conservative than use of HB vehicle.

in the development of the loading model and so reduce the margin of safety. There is perhaps a need for highway and bridge managers to regularly monitor traffic so that they can identify at an early stage any deleterious changes in traffic patterns which might affect the extremes of the loading being carried by a particular structure. If the extreme loadings were such that they were felt to be jeopardising the safety of the structure then some form of remedial action might have to be taken. As this review illustrates most changes and revisions to loading codes have been to catch up with situations which have already developed. It would make sense to develop a more pro-active method of controlling bridge loading to avoid the difficult problems which arise when standards are enhanced and families of existing structures have to be assessed for their carrying capacity, and maybe strengthened or replaced at considerable expense.

References

1. Henderson, W. (1978) *Background to BS 5400: Part 2.* Unpublished report for the Department of Transport.
2. Henderson, W. (1978) *Development of Type HA and Type HB loading.* Unpublished report for the Department of Transport.
3. British Standards Institution (1978) BS 5400: Part 2: 1978, *Steel, concrete and composite bridges, Part 2: Specification for loads.* London: BSI.
4. McMahon, W. and Brady, K. C., Transport Research Laboratory (2000) A *review of the design loads for buried structures.* Unpublished report (2 vols) for the Highways Agency.
5. British Standards Institution (2000) DD ENV 1991-3: 2000, *Eurocode 1: Basis of design and actions on structures—Part 3: Traffic loads on bridges.* London: BSI.
6. Report of Joint Committee of Concrete Institute, Institution of Municipal and County Engineers and Institution of Municipal Engineers (1918) *Loads on highway bridges.* London: The Concrete Institute.
7. Ministry of Transport (1922) *Report on the administration of the Road Fund for the year 1921–1922.* London: HMSO.
8. British Engineering Standards Association (1923) *No. 153—Parts 3, 4 and 5. British standard specification for girder bridges—Part 3: Loads and stresses.* London: Crosby Lockwood & Son.
9. Joint Committee of Institution of Civil Engineers and Institution of Structural Engineers (1949) *Code of practice for simply supported steel bridges.* London: ICE and IStruct.E.
10. British Standards Institution (1954) BS 153: Part 3A: 1954, *Girder bridges—Part 3: Loads and stresses—Section A: Loads.* London: BSI.
11. British Standards Institution (1978) BS 5400: Part 2: 1978, *Steel, concrete and composite bridges—Part 2: Specification for loads.* London: BSI.

Chapter 2
Historical review: Part 1, 1911–54

2.1 Concrete Institute Report 1918 [1]

The development of a national UK standard for traffic loads on highway bridges can be traced back to the setting up of a joint committee in 1911 to look at 'The loads to be provided for upon Highway Bridges'. The committee was composed of representatives from the Concrete Institute, the Institution of Municipal and County Engineers and the Institution of Municipal Engineers. The work of the committee was affected by the 1914–18 war and, although a draft report was submitted in 1912, the final report was not published by the Concrete Institute until 1918.

As part of their studies the committee obtained information from various local authorities about the heaviest vehicles which were used as a basis for the design of road bridges in their particular neighbourhoods. The responses showed that there was no standard approach but that fairly heavy vehicles were catered for with quite high axle and wheel loadings. For instance, the Manchester Ship Canal designed their bridges for a maximum axle load of 30 tons, or 50 tons gross on a 10 ft × 5 ft wheelbase. At Wakefield their bridges were designed for a 15-ton traction engine pulling a 40-ton trolley. In a less industrialised part of the country the Middlesex County Engineer considered that bridges should be designed to carry two 15-ton steam rollers back to back, or two fully laden 12-ton tramcars, with as many people as possible who could crowd onto the remaining part of the carriageway. Examples of heavy loads were also given with one boiler manufacturer quoting a gross load on a trolley of 40 tons, with 25 tons on the rear axle, drawn by a 15-ton tractor with a 10-ton rear axle.

In their review the committee recommended three classes of loading which reflected the importance of the highway concerned. The highways were classified as follows:

Class 1. Bridges on main thoroughfares in great industrial centres and within 25 miles thereof.

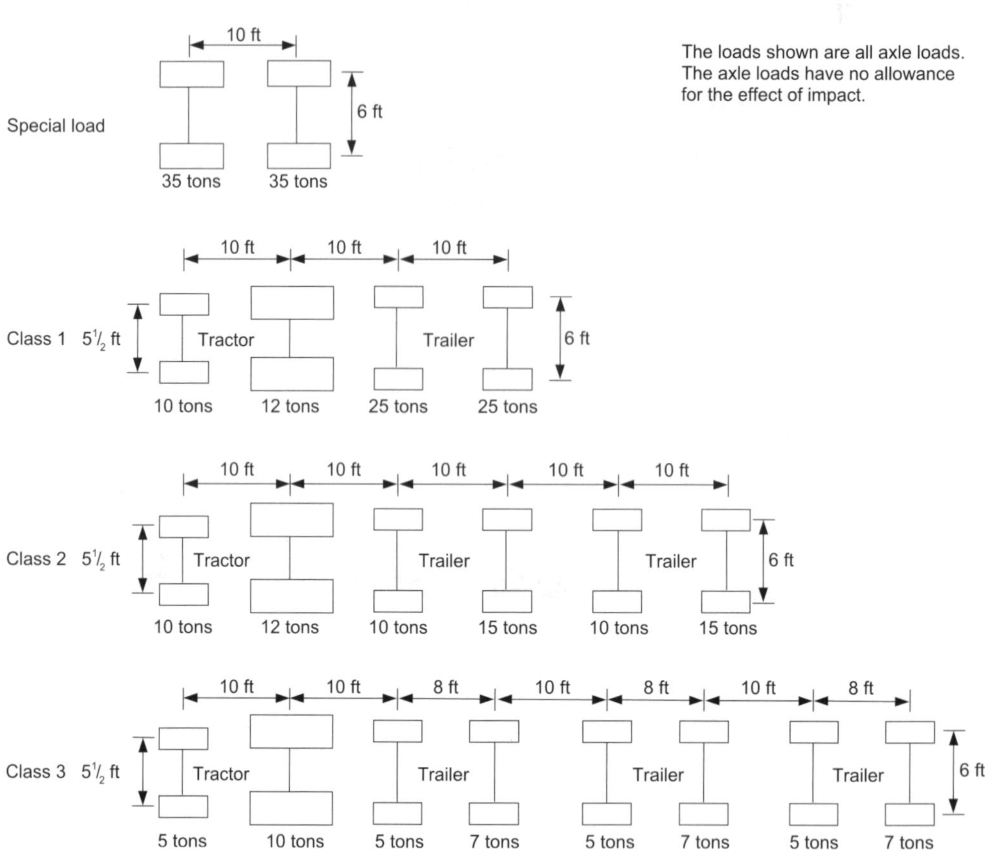

Figure 2.1. Loading trains proposed in 1918 [1]

Class 2. Bridges on secondary roads in great industrial centres and on main roads outside a radius of 25 miles.

Class 3. Bridges on provincial roads other than main roads.

The loadings associated with each class of road took the form of trains with a locomotive drawing a trailer or trailers, as shown in Fig. 2.1. Special loads were specified for particular parts of the country such as the North of England and the Midlands. The trains were assumed to occupy a 10 ft width of road with only one train on the bridge at any one time. The remaining area of the carriageway, including footways, was to carry a distributed load of 1 cwt/ft^2, except for the width of road in front of and behind the load train.

Although clearly aimed at metal bridges with main girders and cross girders, the report did give recommendations for the design of arch bridges. There were recommendations for the distribution of point loads through the thickness of the road material with a spread of 45° in each direction, taking the contact area of the wheel on the road surface as a thin line having a length equal to the width of the tyre. It should be remembered that the wheels of the heavy vehicles specified in the loading trains would be mainly fitted with iron tyres.

The committee were unable to give a general rule for dealing with the dynamic effects of moving loads although they recognised that the condition of the road surface would contribute to the vibration of the load and increase the stresses in the bridge members. However, they suggested that the speed of the vehicles, and hence any impact effects, could be ignored if bridges over 50 ft span were designed with a safety factor of four and bridges under 50 ft span with a safety factor of five, the safety factor being applied to both dead and live loads.

2.2 Ministry of Transport Standard Loading Train, 1922

At the end of the war, in 1918, the Ministry of Transport was established and became responsible for issuing the first national loading rules for bridges. These were contained in the *Report on the administration of the Road Fund for the year 1921–22* [2]. The main live loading consisted of an idealised arrangement of a steam traction engine drawing three trailers as shown in Fig. 2.2, and which became known as the Standard Loading Train (SLT). The train of vehicles was based on the heaviest commonly occurring vehicles such as might be used by agricultural contractors, except that all the actual axle weights were increased by 50 per cent to allow for impact. This might appear to be somewhat excessive, especially in view of the slow speed of such vehicles, but perhaps reflects the lack of knowledge at that time about impact effects. Also the loading rules for small railway bridges at that time required a 50 per cent increment to allow for dynamic effects. If the 50 per cent impact allowance is removed from the SLT there is then reasonable agreement between the SLT and the Joint Committee's Class 3 train except that the rear axles of the locomotive of the former are still almost 50 per cent heavier than those of the latter. The SLT was to be applied to all bridges on principal roads required to carry normal traffic; it was recognised that bridges in certain industrial districts would have to be designed for heavier loads. Thus the intended application of the SLT appears to mirror that of the Joint Committee's Class 3 loading although the definition of road types is not very clear.

The Ministry recognised in the notes to the SLT that a number of different loading situations could occur on the mainly two-lane bridges which were the norm at that time. Trains travelling in opposite directions could pass one another on a bridge and since these trains were slow moving there could be convoys of closely spaced trains. Thus the Ministry's loading rules required an SLT to be placed in every 10 ft width of carriageway with not more than one engine in every 70 ft span of the bridge. In normal cases the distance between the centre lines of adjacent trains was to be taken as 10 ft. Parts of the standard loads could be used if they produced a worse effect than the complete train. The loading appears to have been the same for all spans and there was no minimum headway specified between following trains.

2.3 Ministry of Transport Equivalent Loading Curve, 1931

In 1931 the MoT issued an Equivalent Loading Curve [3] which specified for different loaded lengths the value of the uniformly distributed load which was to be applied to the bridge span or member in question together with an invariable

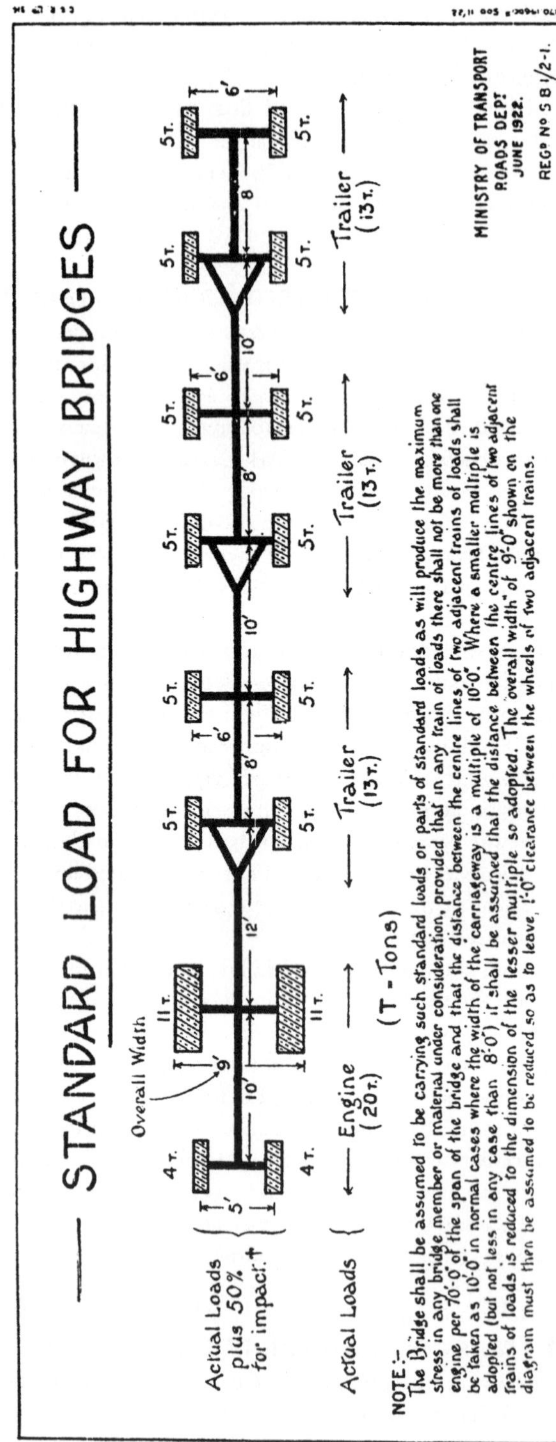

Figure 2.2. Ministry of Transport standard loading train [2]

KEL. The uniformly distributed load was derived from the Standard Loading Train by taking the weights of all the axles of the train on a particular loaded length, assuming all the axles to be 10 tons in weight, and converting them into a uniformly distributed load which, when applied over the whole loaded length, would give the same total weight as the combined weight of all the axles.

The excess weight of the heavy locomotive axle, namely 22 tons minus 10 tons, was taken as a KEL of 2700 lb/ft width which was to be applied across the 10 ft width of the traffic lane. This Equivalent Loading Curve (see Fig. 2.3) was later incorporated in Memorandum No. 577 [4] which was issued by the Ministry of War Transport in 1945 and it remained largely unmodified until the issue of Memorandum No. 771 [5] in 1961. The move from a traffic loading model based on a particular vehicle train to one which was simpler to apply in hand calculations, and which could encompass the loading effects of a variety of vehicles, marked a significant step in highway bridge design. The use of a uniformly distributed load (UDL) together with a KEL has continued to be the format for traffic load models up to the present day. As can be seen from Fig. 2.3, there are two break points in the loading curve; below 10 ft loaded length the UDL rises quite steeply while above 10 ft it remains constant until 75 ft when it reduces to a value of 70 lb/ft^2 at 2500 ft and remains constant from thereon. Notes on the loading in Memorandum No. 577 [4] state that on bridges with spans exceeding 75 ft a reduction has been made in the intensity of loading as compared with the MoT Standard Train. This was said to be to allow partly for a lower average weight of vehicles in large groups, and partly for the reduced effect of impact on the longer spans. There was also a note that for spans below 10 ft the equivalent loading makes allowance for bending moments in both directions, of which only the main bending moment may be calculated.

It has been noted that the Equivalent Loading Curve model was derived by transforming the weights of the individual axles in the standard loading train into an equivalent variable UDL and an invariant KEL which, together, for each span, produced the same total weight on the structural element as the total of the axle weights. A more accurate and relevant method would have been to have derived the equivalent loading by comparing the bending moments and shearing forces produced by the loading train and the load model respectively at critical locations. As far as the structure is concerned it is the load effects on individual structural members that are important, rather than the total weight carried by the structure or member. This total weight method of deriving the equivalent loading can result in calculated load effects which underestimate the actual load effects produced by the axles of the loading train. As will be seen when looking at later versions of equivalent loading curves it was somewhat fortuitous that the 1922 Standard Loading Train included the very high 50 per cent impact factor on all the axles.

2.4 BS 153—*Standard specification for girder bridges*, 1923

While the MoT had been developing its loading requirements for highway bridges, the British Engineering Standards Association (later to become the British Standards Institution (BSI)) had been developing a comprehensive specification for the design

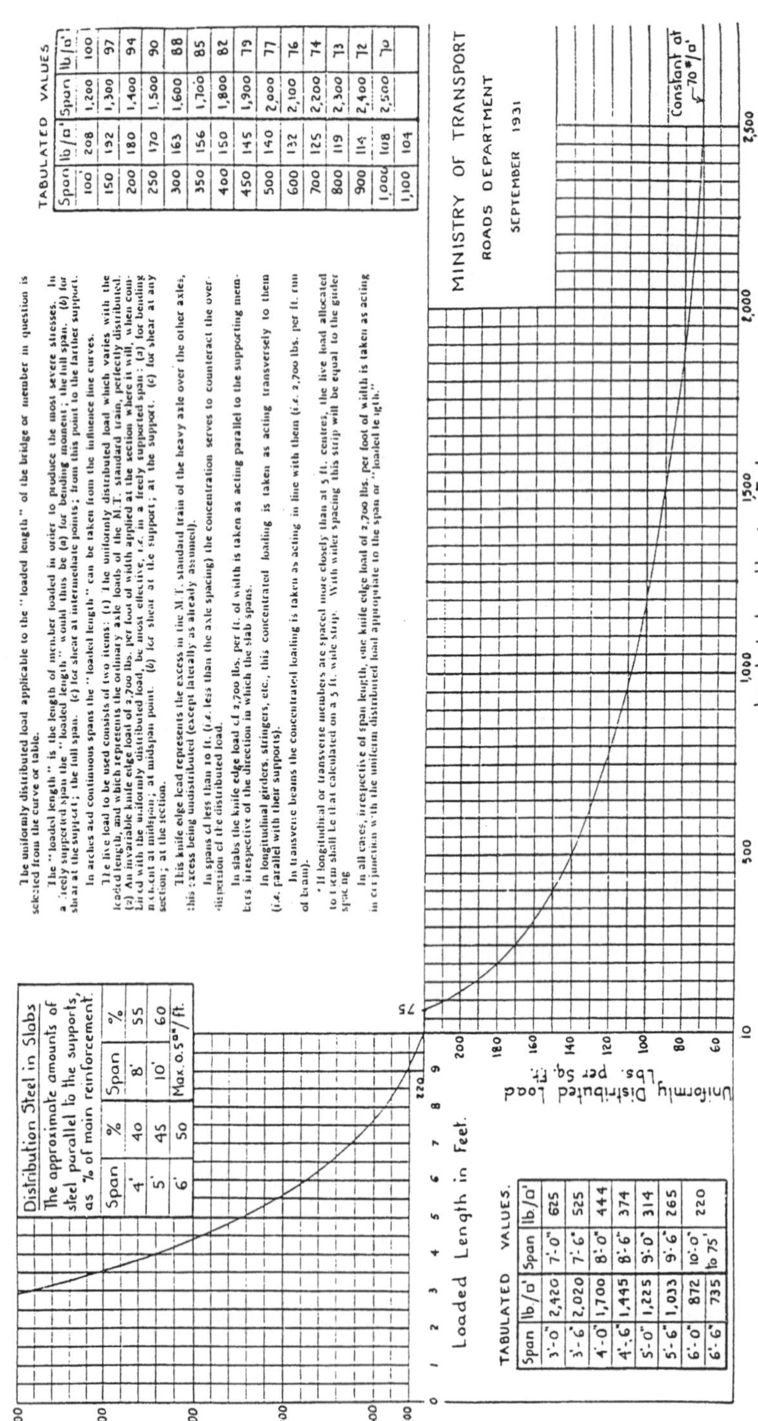

Distribution Steel in Slabs

The approximate amounts of steel parallel to the supports, as % of main reinforcement.

Span	%	Span	%
4'	40	8'	55
5'	45	10'	60
6'	50	Max. 0.5□"/ft.	

TABULATED VALUES.

Span	lb./□'	Span	lb./□'
3'-0"	2,420	7'-0"	625
3'-6"	2,020	7'-6"	525
4'-0"	1,700	8'-0"	444
4'-6"	1,445	8'-6"	374
5'-0"	1,225	9'-0"	314
5'-6"	1,033	9'-6"	265
6'-0"	872	10'-0" to 75'	220
6'-6"	735		

The uniformly distributed load applicable to the "loaded length" of the bridge or member in question is selected from the curve or table.

The "loaded length" is the length of member loaded in order to produce the most severe stresses. In a freely supported span the "loaded length" would thus be (*a*) for bending moment; the full span. (*b*) for shear at the support; the full span. (*c*) for shear at intermediate points; from this point to the farther support.

In arches and continuous spans the "loaded length" can be taken from the influence line curves.

The live load to be used consists of two items: (1) The uniformly distributed load which varies with the loaded length, and which represents the ordinary axle loads of the M.T. standard train, perfectly distributed. (2) An invariable knife edge load of 2,700 lbs. per foot of width applied at the section where it will, when combined with the uniformly distributed load, be most effective, *i.e.* in a freely supported span: (*a*) for bending moment at midspan; at midspan point. (*b*) for shear at the support; at the support. (*c*) for shear at any section; at the section.

This knife edge load represents the excess in the M.T. standard train of the heavy axle over the other axles, this excess being undistributed (except laterally as already assumed).

In spans of less than 10 ft. (*i.e.* less than the axle spacing) the concentration serves to counteract the over-dispersion of the distributed load.

In slabs the knife edge load of 2,700 lbs. per ft. of width is taken as acting parallel to the supporting members irrespective of the direction in which the slab spans.

In longitudinal girders, stringers, etc., this concentrated loading is taken as acting transversely to them (*i.e.* parallel with their supports).

In transverse beams the concentrated loading is taken as acting in line with them (*i.e.* 2,700 lbs. per ft. run of beam).

* If longitudinal or transverse members are spaced more closely than at 5 ft. centres, the live load allocated to them shall be that calculated on a 5 ft. wide strip. With wider spacing this strip will be equal to the girder spacing

In all cases, irrespective of span length, one knife edge load of 2,700 lbs. per foot of width is taken as acting in conjunction with the uniform distributed load appropriate to the span or "loaded length."

TABULATED VALUES

Span	lb/□'	Span	lb/□'
100	208	1,200	100
150	192	1,300	97
200	180	1,400	94
250	170	1,500	90
300	163	1,600	88
350	156	1,700	85
400	150	1,800	82
450	145	1,900	79
500	140	2,000	77
600	132	2,100	76
700	125	2,200	74
800	119	2,300	73
900	114	2,400	72
1,000	108	2,500	70
1,100	104		

Figure 2.3. MoT Equivalent Loading Curve [3]

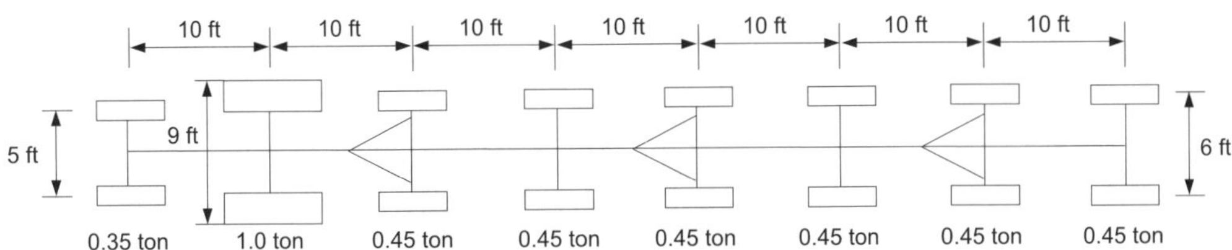

Note: The impact allowance to be added to the above loads varied with the span of the structure

Figure 2.4. British Standard Unit Loading for Highway Bridges [6]

and construction of steel girder bridges. Part 3 of the document [6], later to be known as BS 153, covered loads and stresses and was issued in 1923, with an appendix, which gave details of the so-called *British standard unit loading for highway bridges*, being issued in 1925. In addition to details of the loading train shown in Fig. 2.4, the appendix also tabulated the bending moments and shears on main girders and longitudinal bearers for spans up to 300 ft together with cross-girder reactions. The presentation of these effects clearly reflected the method of construction in use at that time with longitudinal road bearers being supported by cross-girders spanning between main girders.

The loading train was similar, in terms of numbers and spacing of axles, to the MoT Standard Loading Train but, unlike the latter, the loading was specified in terms of unit tons per axle, the heaviest axle being 1 ton and the rest 0.45 tons. The designer was expected to select the appropriate number of units of loading to be applied; each 10 ft width of carriageway was to be completely occupied by a succession of standard loading trains with a 10 ft headway between each train. It was noted that the MoT recommended that a minimum of 15 units should be used in the design of all road bridges in Great Britain. In addition to the loading trains a UDL of 84 lb/ft^2 could be applied to areas carrying pedestrian or equivalent light traffic. The impact effect was calculated from a simple formula which took account of the loaded length of the member concerned and the number of lines of traffic supported by that member with a maximum value of 0.7.

2.5 BS 153—*Standard specification for girder bridges*—1937 revision [7]

Part 3 of BS 153 was reissued in 1937 [7], the main change, as far as traffic loading was concerned, being the introduction of a Type B loading, as shown in Fig. 2.5. This was primarily intended for constructions overseas and was to be applied to one lane only, although the lane was to be completely occupied by a succession of the standard trains which were assumed to occupy a width of 9 ft. The length of the trailers in the Type B train varied according to the number of units of loading being considered but the dimensions of the bogies themselves were fixed. An alternative two-axle vehicle for heavy loading was also specified. The original Standard Unit Loading was named Type A loading; 15 units of the standard loading were still accepted by the MoT for the design of highway bridges.

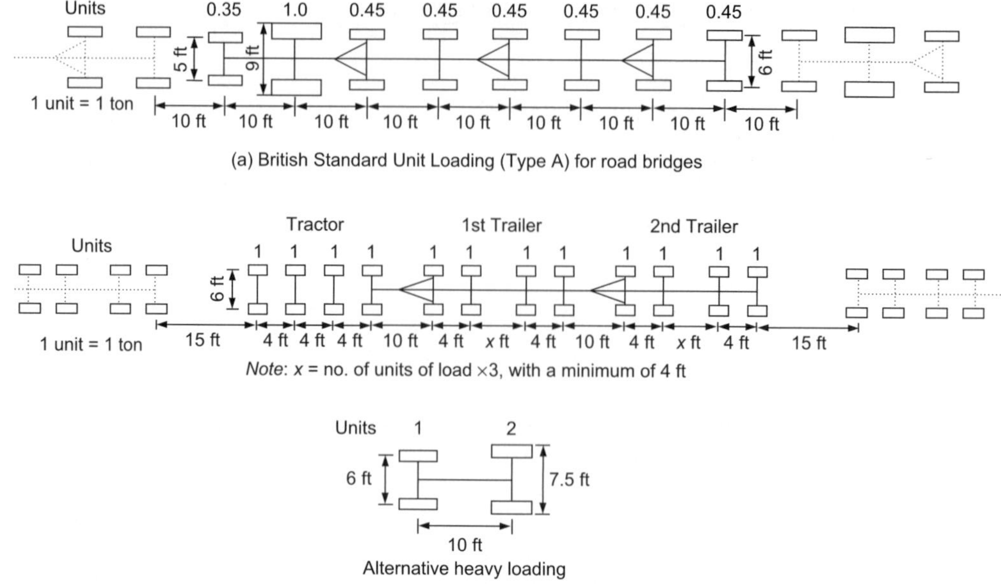

Figure 2.5. British Standard Unit Loading (BS 153, 1937) [7]

New rules for assessing impact were introduced with the impact factor reducing with increasing span. For bridges carrying one lane of traffic the maximum impact factor was 60 per cent whilst for bridges carrying two or more lanes it was 50 per cent. It was assumed that the surface of the carriageway on and immediately adjacent to the bridge was maintained in good condition.

2.6 *Code of practice for simply supported steel bridges, 1949 [8]*

In 1936 a joint committee was formed by the Institutions of Civil and Structural Engineers to formulate a code of practice for simply supported steel bridges. The work of the committee was suspended during the Second World War but was resumed in 1946 with the code of practice being published in 1949 [8]. The code of practice covered both rail and road bridges with spans up to 300 ft.

Although the code of practice introduced a number of innovations for traffic loading, it retained the MoT Equivalent Loading Curve which was to be applied, along with the 2700 lb/ft KEL, to a 10 ft width of carriageway. It was noted that this loading corresponded approximately to 20 units of the British Standard Unit Loading with appropriate impact and dispersion allowances. Both the Type A and Type B British Standard Unit Loadings were included. In addition a new Abnormal Loading, known as Type C, was introduced to represent the increasing use of multi-axle trailers to transport the heavy loads, such as electricity generators, which were required in the post-war reconstruction. This Type C load was to be applied in

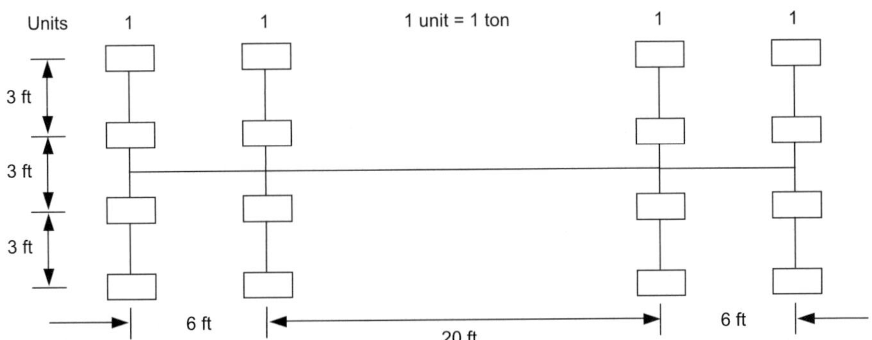

Figure 2.6. ICE Code of Practice Type C Loading (1949) [8]

addition to, but independent of, the Type A load, the Type B load or the Ministry of Transport Standard Load.

The plan and axle arrangements of the Type C vehicle are shown in Fig. 2.6, where each of the four axles have a designated loading of one unit or 1 ton. In Great Britain 45 units of the Type C loading were to be used but it was permissible to overstress steel members by 25 per cent. The *Code of practice* gave simplified rules for calculating the effects of the Type C loading on longitudinal and transverse members; slabs were to be designed to carry the MoT Standard Loading but with certain provisos. Where Type C loading was considered to be appropriate, such as on the designated heavy load routes, one train of that type was to be assumed to occupy two traffic lanes for the whole length of the bridge. Type C loading with increased permissible stresses was suggested as an alternative to the Ministry of Transport Standard Loading at normal permissible stresses with the worst case being taken.

Among the other innovations was the allowance for a reduction in loading on wide bridges, so that on bridges with more than two lanes only two lanes were to be occupied by the full loading with the remaining lanes being occupied by half of the full loading. The committee had received considerable evidence that the effect of impact was considerably less than had been previously allowed and therefore recommended a radical reduction in the allowance in cases where the Ministry of Transport Standard Loading, which included a 50 per cent allowance, was not being used. For slabs, stringers and girders of less than 100 ft span an impact allowance of 50 per cent was to be added only to the individual concentrated load producing the greatest bending moment or shear. For spans greater than 100 ft impact could be ignored.

The *Code of practice* also introduced specifications for the longitudinal forces resulting from the braking of vehicles. These were to be taken as either 50 per cent of the load from the first two vehicles of the British Standard Unit Loading Types A and B, 37.5 per cent of the live load from the first 30 ft of the Ministry of Transport Standard Loading or 20 per cent of the total weight of the Type C abnormal vehicle. Although requirements for calculating the centrifugal forces on railway bridges were given it was not considered necessary to calculate such forces for road bridges. The *Code of practice*, with some revisions, formed the basis of a later revision of BS 153 (1937).

2.7 Ministry of War Transport Memorandum No. 577, 1945 [4]

It will have been noted that the standards and codes considered so far were primarily intended for the design of steel bridges, although there were some references to the loads on concrete slabs. Many of the loading criteria were specially tailored for particular aspects of the then current steel bridge designs such as cross-girders and stringers. The Ministry of War Transport filled this gap in the materials covered by issuing Memorandum No. 577 in 1945. This covered all aspects of the design of highway structures in steel, concrete and mixed steel and concrete as well as the use of prestressed concrete beams. It is interesting to note that this document, which also covered ancillary components such as parapets, expansion joints and bearings, consisted of only 50 A5 pages, 30 of which were diagrams and figures.

All bridges and culverts were to be designed or strengthened to carry the Ministry of War Transport Standard Loading as specified in the form of the 1931 Equivalent Loading Curve which was attached to the Memorandum. The Memorandum also gave requirements for the loading on footpaths and cycle tracks where, as an alternative to the application of 100 lb/ft^2, a single 4-ton wheel load could be considered. For the design of slabs supported on all four sides a single 20-ton wheel load was specified. The MoT's requirements for loading in Memorandum No. 577 remained in force until it was superseded by Memorandum No. 771 in 1961 [5].

References

1. Report of Joint Committee of Concrete Institute, Institution of Municipal and County Engineers and Institution of Municipal Engineers (1918) *Loads on highway bridges*. London: The Concrete Institute.
2. Ministry of Transport (1922) *Report on the administration of the Road Fund for the year 1921–1922*. London: HMSO.
3. Ministry of Transport (1931) *Standard load for highway bridges (single sheet)*. London: Ministry of Transport.
4. Ministry of War Transport (1945) Memorandum No. 577, *Memorandum on bridge design and construction*. London: HMSO.
5. Ministry of Transport (1961) Memorandum No. 771, *Standard highway loadings* (*amendment to Memorandum No. 577*). London: HMSO.
6. British Engineering Standards Association (1923) No. 153—Parts 3, 4 and 5—1923. *British standard specification for girder bridges—Part 3: Loads and stresses*. London: Crosby Lockwood & Son.
7. British Standards Institution (1937) No. 153, *British standard specification for girder bridges—Part 3: Loads and stresses*. London, BSI.
8. Joint Committee of Institution of Civil Engineers and Institution of Structural Engineers (1949) *Code of practice for simply supported steel bridges*. London: ICE and IStruct.E.

Chapter 3
Historical review: Part 2, 1954–78

3.1 BS 153: Part 3A: 1954 [1]

Much of the background to this revision of the loading specification in BS 153 was given in a paper by Henderson to the Institution of Civil Engineers in March 1954 [2]. In preparing for a second revision of BS 153: Part 3A two areas of concern about the applicability of the traffic load models were identified. First, there had been a general increase in the weight and number of lorries with the maximum weight under the *Construction and use regulations* being raised to 22 tons. Second, there had been an increase in the number and weight of the special vehicles carrying abnormal indivisible loads which were governed by the *Authorisation of special types—general order* [3] regulations. These were limited to 150 tons maximum but it was possible to obtain a special order to move heavier loads. The main concern was that the vehicle trains used to develop the various loading curves and loading requirements were not fully representative (both in weight and axle/wheel configurations) of the ordinary and abnormal traffic of the day. Moreover, the original loading trains had been based on solid tyred wheels whereas, even for abnormal vehicles, pneumatic tyres with greater contact areas and wider spacing of the wheels on axles were in everyday use. It was therefore decided to re-evaluate the basis of both the normal and the abnormal loading requirements with the aim that both the MoT and the BSI should be able to adopt the single set of loading rules. In the case of the normal loading it was felt that the adoption of a uniformly distributed load with an associated KEL would make computation easier and find favour with designers.

3.1.1 Loading models

The development of an equivalent loading model has to take account of both bending moments and shears. If only UDLs by themselves are adopted to represent the vehicle effects then two separate UDL models would be required, one for bending moments and one for shear. The use of a *point* load such as the KEL, placed in a position where it will have the greatest effect, makes it possible to use just one

UDL model and obtain quite good correspondence with the vehicle load effects for both bending moment and shear. Strictly speaking, to obtain exact equivalence would require the use of two KELs, one for moment and one for shear, but this would make design more complicated. Thus for simplicity of design, and bearing in mind that traffic load modelling can never be an exact science, it is generally considered that the use of one loading model (i.e. UDL + KEL) for both bending moment and shear calculations is justified. For longer loaded lengths, where the large number of individual axle loads becomes more like a uniformly distributed load, any discrepancies will be minimised. The KEL is thus a device to obtain an equivalence between the loading effects obtained from actual vehicles and those obtained from the loading model, and it should not necessarily be taken as representing the heaviest axle in a train of vehicles.

3.1.2 Normal loading curve

In developing the normal loading it was decided to investigate whether loading criteria could be formulated which, while being adequate for the then current traffic, would nevertheless not differ widely from the MoT Equivalent Loading Curve. Two different trains of vehicles were used for the comparisons for spans up to 75 ft and spans of 75–500 ft respectively, as shown in Fig. 3.1. A 4-axle 22-ton vehicle with

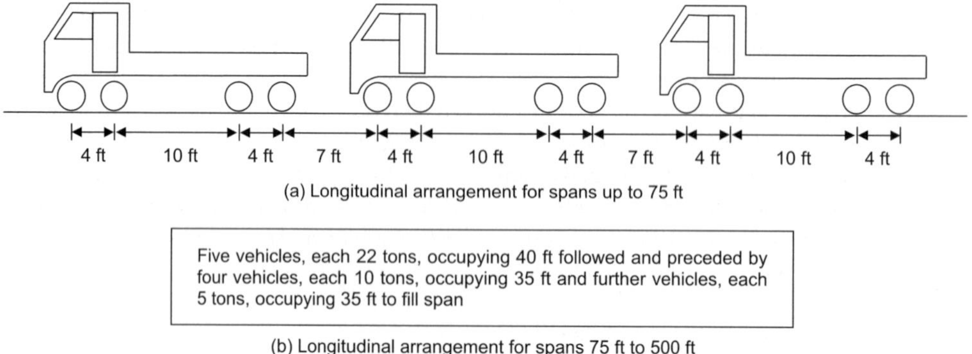

(a) Longitudinal arrangement for spans up to 75 ft

Five vehicles, each 22 tons, occupying 40 ft followed and preceded by four vehicles, each 10 tons, occupying 35 ft and further vehicles, each 5 tons, occupying 35 ft to fill span

(b) Longitudinal arrangement for spans 75 ft to 500 ft

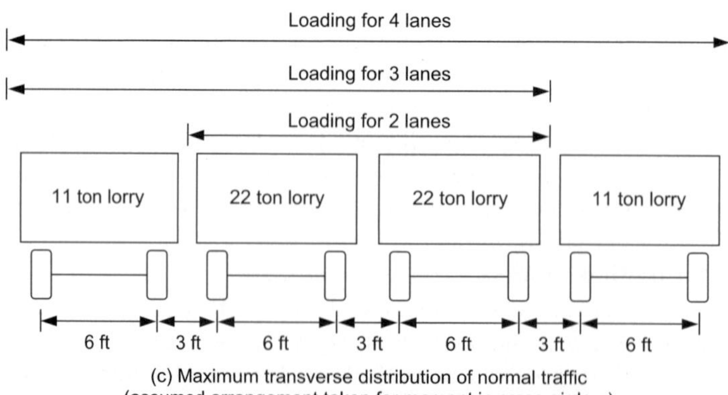

(c) Maximum transverse distribution of normal traffic
(assumed arrangement taken for moment in cross-girders)

Figure 3.1. Vehicle arrangement used to derive HA loading (after [2])

a short wheelbase was chosen as the main vehicle in the train because it provided a particularly onerous loading. The comparisons were based on the loading effects (i.e. bending moments and shears) for longitudinal and transverse members. For spans between 20 and 75 ft the comparisons showed that effects from the vehicle trains were very close to those derived from the Equivalent Loading Curve and thus it was felt that the invariable uniformly distributed load could be retained for that span range.

The choice of the dividing point between 'short' loaded length and 'long' loaded length was somewhat arbitrary but concurred with the step in the Equivalent Loading Curve. It was felt that above 75 ft span vehicles in one lane would be in random positions widthwise and thus only some of the wheels in a train of vehicles would be running directly over a longitudinal girder. It was also felt that vehicles in one lane would not coincide longitudinally with vehicles in an adjacent lane. Between 75 and 500 ft span the comparison of loading effects was therefore based on a single line of vehicles and again a uniformly distributed form of loading was derived which, with a 27 000 lb KEL, could replicate the loading effects from the vehicle train for different spans. Some adjustments were needed at the 75 ft span to marry in the straight line portion of the loading with the curved part. From 500 to 3000 ft span the uniformly distributed load was calculated assuming an average weight of traffic per linear foot of lane; this represented a train of the heaviest vehicles with a private car between each truck. This assumption equated to an average gap between the vehicles of the day of about 55 ft; using a 2-second headway rule this meant that the vehicles would be travelling at about 18 mph. It was felt that this concentration of heavy traffic could probably occur in one lane but was unlikely to occur simultaneously in two adjacent lanes. An allowance for impact was included in the derivation of the normal loading from the train of vehicles. This amounted to a 25 per cent increase in the load of any one axle of one vehicle, or any single pair of adjacent wheels of two vehicles travelling abreast. The nominal loading curves which were derived and eventually included in BS 153 are shown in Fig. 3.2.

3.1.3 Short-span members

For short-span members and units supporting small areas of deck it was felt desirable to specify a heavier concentrated load than that provided by the single wheel or axle of the 22-ton lorry. Based on the weights of the heavy wheels in the earlier vehicle trains, say about 7 tons, and adding a 25 per cent impact allowance gave a design load of 9 tons. For the purposes of calculating the equivalent loading for these small units two of these 9-ton wheels set 3 ft apart, representing the wheels of two adjacent vehicles, were used to calculate the equivalent uniformly distributed loading for the 3–20 ft span range. Again the derived curve had to be adjusted to marry up with the horizontal portion at 20 ft span. The loading curve, which had been derived for longitudinal members, was used in conjunction with the KEL to calculate the loading effects on cross-girders carrying two, three and four lanes of traffic. It was found that the effects calculated from the vehicles were about 25 per cent lower than those calculated from the proposed design loading, but this was felt to be acceptable allowing for possible impact effects and greater concentrated loads.

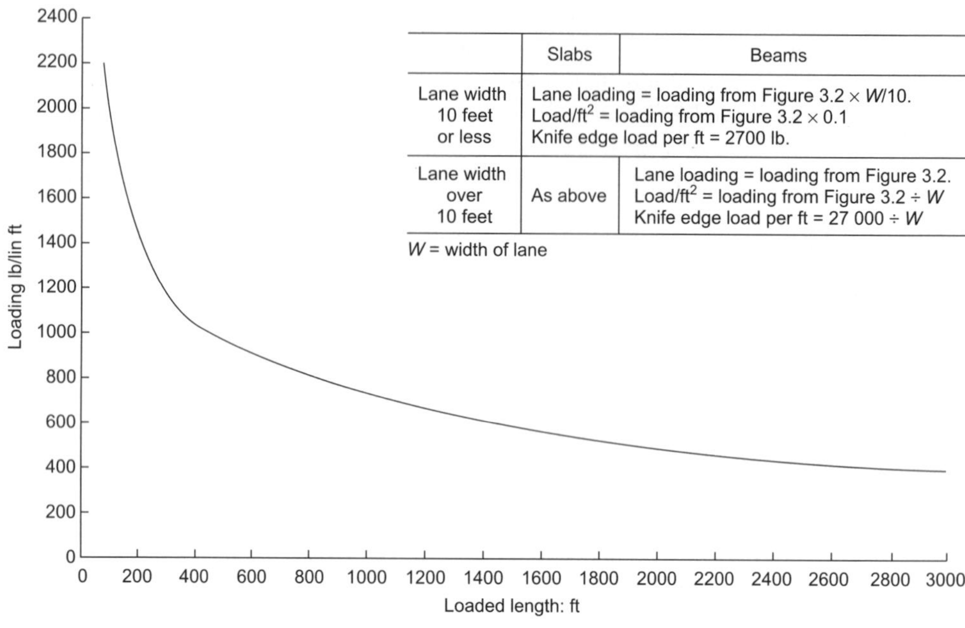

	Slabs	Beams
Lane width 10 feet or less	Lane loading = loading from Figure 3.2 × *W*/10. Load/ft^2 = loading from Figure 3.2 × 0.1 Knife edge load per ft = 2700 lb.	
Lane width over 10 feet	As above	Lane loading = loading from Figure 3.2. Load/ft^2 = loading from Figure 3.2 ÷ *W* Knife edge load per ft = 27 000 ÷ *W*

W = width of lane

Figure 3.2. Loading curve for Type HA highway loading [1]

3.1.4 Concrete slabs

An investigation of reinforced concrete slabs using the train of vehicles showed that they could be under-designed using the proposed design loading. It was therefore decided that such slabs should be designed either for a two-axle bogie weighing 60 tons, each axle having four wheels, or for two 11.25-ton wheels. These loads were then used to derive the equivalent UDLs to be applied to longitudinal and transverse slabs with the invariant KEL; these are shown in Fig. 3.3. It will be noted that the slab loading does not extend beyond the 20 ft span as the effects from the train of C&U vehicles then becomes more severe than the effects from the special bogies.

3.1.5 Notional traffic lanes

The original MoT loading trains and equivalent loading curve were based on the premise that convoys of vehicles would occupy a 10 ft width of the carriageway and the early road lane markings adopted this figure. Shortly after the end of the second world war this was increased to 11 ft, although 12 ft was adopted for curves. Consideration of the actual width of vehicles and of how they might actually arrange themselves on a bridge, especially at times of congestion, led to the realisation that there was a need to rationalise the way that the design loading was applied to the bridge deck. This led to the introduction of the concept of dividing the carriageway into notional lanes for loading purposes, the lanes not necessarily coinciding with the lane markings on the road surface. For BS 153 it was decided that full lane loading would be applied if the width of the notional lane was

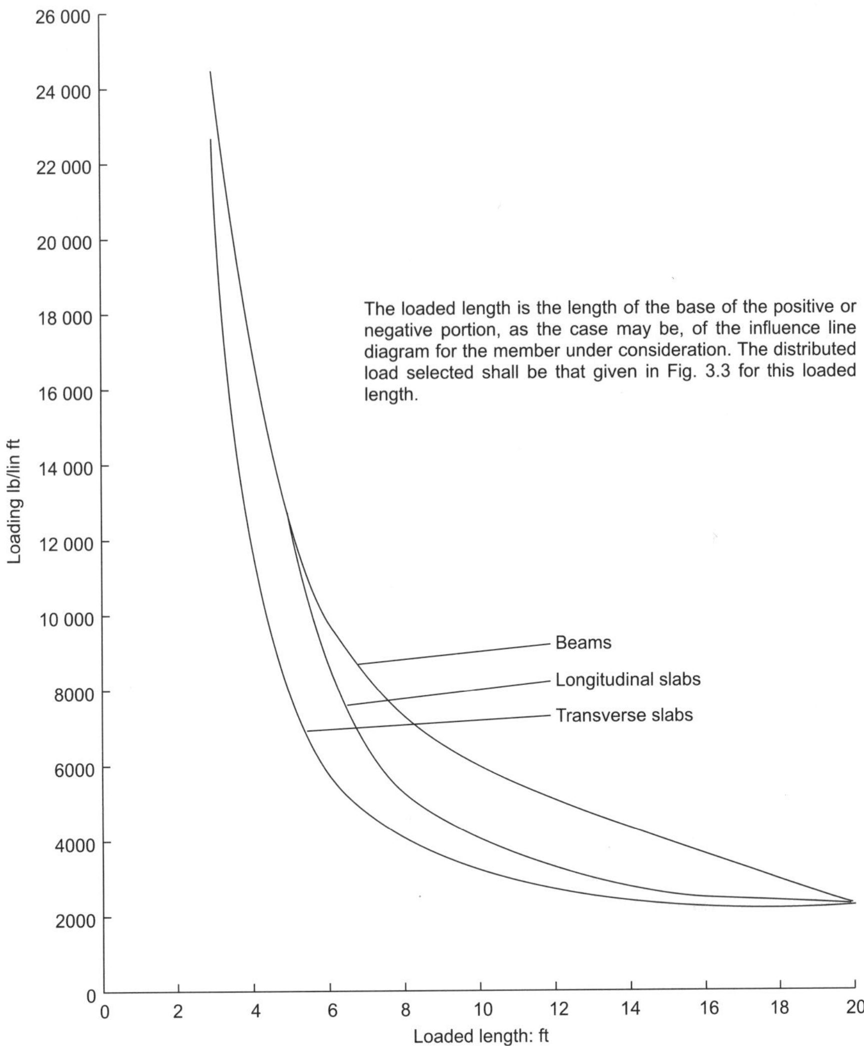

Figure 3.3. HA loading for short spans [1]

between 10 ft and 12 ft. Where notional lanes were less than 10 ft wide there could be a pro-rata reduction in the lane loading, as it was assumed that the vehicles occupying such lanes would be narrower and hence lighter than the vehicles used to generate the UDL. With the introduction of notional lanes the UDL was presented as the load per unit loaded length which was to be applied, together with the KEL, across the full width of a notional lane between 10 ft and 12 ft wide. Although during the development of BS 153 it was proposed that if there were more than two lanes on a bridge then the additional lanes should be loaded with half the full lane loading, this was reduced to one third after further consideration.

3.1.6 Abnormal loading

The loading described above for normal vehicles was designated Type HA loading; it was decided that there was also a need for a loading to cover the heavy trailers which were increasingly being used to transport very heavy abnormal loads around the UK. This loading was subsequently designated Type HB loading. The MoT had been getting concerned about the possible severe effects on bridges of some of the vehicles being used; one trailer in particular carried 200 tons on two bogies each with two closely spaced axles. Discussions between the Ministry and industry led to the adoption of a maximum weight of 150 tons for such trailers. At the same time the Ministry was preparing legislation to control the usage of such vehicles by limiting the weights and spacings of the wheels and axles. The Statutory Instrument containing the regulations [3], although fixing the maximum permissible trailer weight as 150 tons, did allow up to 90 tons to be carried on a bogie provided there was a 6 ft spacing between the axles.

In developing the model vehicle that was to be used in design it was decided to increase the weight of the trailer to 180 tons to allow for the effects of the two tractors which would have pulled and pushed the heavy trailer; the laden weight of such tractors at the time was about 17.5 tons. Thus the HB design vehicle was given a maximum weight of 180 tons carried on two double-axle bogies with the minimum spacing between the axles being that laid down in the regulations. The HB vehicle was intended to represent a whole range of abnormal vehicles with the 180 ton vehicle with 45-ton axles being the severest loading. The arrangement of wheels and axles was identical to that of the Type C vehicle in the ICE *Code of practice* 1949, as shown in Fig. 2.6.

The HB vehicle was shown in BS 153 with 1 ton units of loading on each axle. Roads carrying the heaviest class of load were to be designed for 45 units of HB loading; other less important roads could presumably have been designed for 20 or 30 units of loading though no advice was given. The HB vehicle was to be applied to only one notional lane with the remainder being occupied by one third of the HA loading. No impact factors were to be applied since it was assumed that these were slow moving vehicles. Special rules were given for the loading of reinforced concrete slabs where variations of HA loading equivalent to 45 units of HB loading could be used for convenience. Bridges on heavy load routes were expected to be designed for HA loading initially and then checked for the appropriate number of HB units of loading. The contact area for the wheel of a 45 unit HB vehicle, each wheel load being 11.25 tons, was specified as 15 in × 3 in. This was based on the observed footprint of a solid tyred wheel which was the common form of wheel fitted to heavy-load trailers at that time.

3.1.7 Braking forces

Another innovation in BS 153 was the introduction of braking forces to be used in the design of highway bridges. These were referred to as longitudinal forces, covering both braking and traction, although braking was the dominant force. Only one such force was to be considered for a bridge: for HA loading this was a force of 10 tons for spans up to 10 ft increasing with span up to a maximum of 25 tons. For HB loading a force of 45 tons was specified for all spans.

3.1.8 Footways

A loading of 80 lb/ft^2 was to be applied on footways of spans up to 75 ft; above 75 ft the footway loading was based on the relevant uniformly distributed load reduced by a factor. The footway was also expected to carry a wheel load of 4 tons.

Three amendments to BS 153: Part 3 were issued in 1959, 1961 and 1966 respectively and these were later incorporated in a reset and reprinted version which was issued in 1966.

3.2 Ministry of Transport Memorandum No. 771, 1961 [4]

The use of BS 153: Part 3A: 1954 was adopted for the design of both steel and concrete highway bridges by the MoT in 1961. Memorandum No. 771 [4] dealt with the number of units of HB loading which were to be applied to roads other than those carrying the heaviest loads. Trunk roads and motorways were to be checked for 45 units, Class I and Class II roads checked for 37.5 units and Class III were to be designed for HA loading only. There was also a proviso that the Ministry should be consulted about the reduced lane loading which was to be applied to bridges with more than two lanes. It appears that this requirement may have arisen because of the lack of clarity in the definition of a carriageway in BS 153. With the construction of bridges carrying dual carriageways the Ministry felt that the BS 153 rule might be applied to each half of the dual carriageway separately when sorting out the number of notional lanes, rather than considering both carriageways together. There had also been some doubt whether the HA loading based on 22-ton vehicles would cover a proposed tanker vehicle weighing 28 tons; in the event it was found that the HA loading would be more than adequate.

Because Memorandum 771 [4] covered concrete as well as steel structures, it dealt with an anomaly concerning the allowance of 25 per cent overstress under HB loading when designing pre-stressed concrete bridges. The Ministry required there to be no bending tensile stresses at working loads and so for the design of pre-stressed concrete slabs a 9-ton wheel load with no provision for overstress was to be used instead of an 11.25-ton wheel with overstress. There were also other requirements concerning the loading for the design of cantilever slabs. Memorandum No. 771 also drew attention to the possible need to consider the effects of centrifugal forces on highway bridges curved in plan.

3.3 Ministry of Transport Interim Memorandum (bridges) IM 10, 1970 [5]

Interim Memorandum 10 was issued in 1970 to deal with queries regarding the interpretation of BS 153: Part 3A and Memorandum No. 771. In particular IM 10 gave the background to the rule in BS 153 which allowed a variation of HA loading to be used for the design of reinforced concrete slabs instead of 45 units of HB loading. IM 10 also gave guidance on the application of highway loading to the design of earth retaining structures and short-span buried precast segmental structures such as culverts.

Of some significance for the design of highway structures was the recognition of the economic advantages to be gained by using some form of load distribution

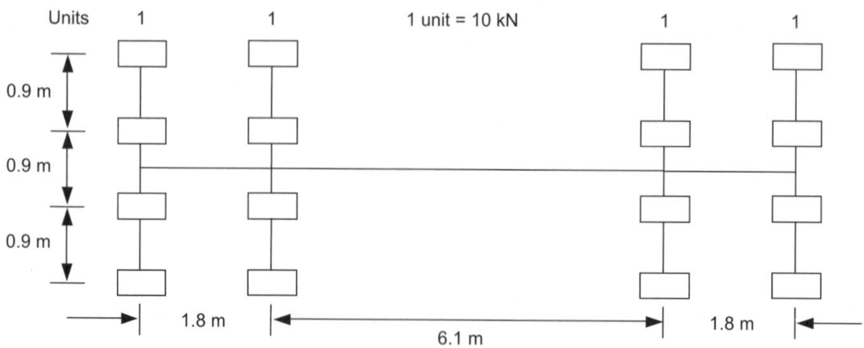

Figure 3.4. Arrangement of HB vehicle [6]

analysis. As well as certain approved hand methods for the analysis of slabs and grillages, IM 10 also drew attention to the increasing availability of computer packages being made available by the Ministry. The more sophisticated methods of analysis had a big advantage in that they could determine directly the load effects in different elements of a structure without the need to use rules of thumb, as had been the practice in earlier codes for dealing with, for example, edge effects.

3.4 BS 153: Part 3A: 1972 [6]

This edition of BS 153: Part 3A, issued in 1972, was a metric edition of the parts of the 1954 version which covered highway loads, although imperial units were still retained for railway bridges. The reason given was that the relatively small demand for railway loading criteria at the time didn't justify the extensive amount of reworking and reprinting that would be necessary to convert the railway loading to metric units. For highway bridges the unit load of the HB vehicle axle, which is shown in Fig. 3.4, was given as 10 kN and there was a slight increase in the equivalent weight of individual wheels from 112.1 kN to 112.5 kN.

The 1972 version also included the earlier amendments to the 1954 version which had been incorporated in the 1966 version. In particular it dealt with the ambiguity concerning the allocation of loading to the notional lanes on a single superstructure carrying dual carriageways which had been noted in Memorandum No. 771. Amendments to Part 3A were probably limited because of the preparation of a new and more comprehensive British Standard for the design and construction of bridges.

3.5 Department of the Environment Technical Memorandum (bridges) BE 5/73 [7]

Technical Memoradum BE 5 was issued by the Department of the Environment, which now covered transport, in 1973 and superseded Memorandum No. 771 [4] and Interim Memorandum IM 10. It was to be used in conjunction with BS 153: Part 3: 1972 [6] for the design of all structures on motorways and trunk roads, except for steel box girder bridges which were to be designed using the *Interim design and workmanship rules* produced by the Merrison Committee [8]. Although the Department was assisting the BSI in the preparation of a comprehensive bridge design code based on limit state principles, BE 5 retained the permissible stress approach with some increases in basic allowable stresses for certain loading combinations including HB loading.

In addition to the requirements for bridges on motorways and trunk roads and principal roads to be checked for 45 and 37.5 units of HB loading respectively, BE 5 required bridges on other public roads to be checked for 30 units of HB loading. Accommodation bridges and bridleway bridges were to be designed for HA loading only. BE 5 also recognised the advances that were being made in bridge construction by specifying the method of applying loading to multi-level structures and to superstructures carrying eight or more traffic lanes.

Technical Memoradum BE 5 introduced the concept of accidental wheel loading to be applied to areas such as a footway which might be vulnerable if an errant vehicle left the carriageway. The loading consisted of three pairs of wheels spaced 1.8 m apart which were meant to represent a lorry and replaced the requirement in BS 153: Part 3A for a single-wheel load to be considered. BE 5 also contained requirements for the effects of vehicle collisions with bridge parapets to be considered along with an impact loading on bridge supports. The design criteria for the former were dealt with in another Departmental Technical Memorandum but the aim of the BE 5 requirements was to ensure that the bridge deck was strong enough to withstand any forces from the parapet supports generated as a result of an accident. The bridge supports were to be protected by a safety fence but were also to be strong enough to resist a horizontal force of 225 kN at right angles to the carriageway and one of 75 kN parallel to the carriageway. In addition a residual force of 150 kN in both directions above the level of the safety rail had to be taken into account.

A major change in BE 5 was to the HA loading curve where for loaded lengths up to 6.5 m (20 ft) the value of the UDL was to taken as a constant 31.5 kN/m, thus extending the already existing horizontal portion of the curve. It appears that because the derivation of the loading curve up to 6.5 m was dominated by the two 112 kN (11.5 ton) wheel loads it was felt that the design of short spans would be adequately covered by either the requirement to design for a minimum of thirty units of HB loading or, if there was no HB requirement, the need to consider two 112 kN wheel loads anyway.

In BE 5 there was also a change in the way that the contact area under HA and HB wheel loads was to be specified. The wheel load in question was to be assumed to be uniformly distributed over a circular or a square area with an effective pressure of 1.4 N/mm^2. This was a considerable reduction on the pressure of 4.0 N/mm^2 which would have been obtained when considering the pressure under a 45 unit HB wheel load based on the 375 mm $\times$ 75 mm contact area. The lower pressure was a recognition of the fact that pneumatic tyres had generally replaced solid tyres on the heavy load trailers. Technical Memoradum BE 5 contained a reminder about the use of load distribution methods of analysis for determining bending moments and shear forces using either recognised hand methods or computerised frame and finite element methods. There were also requirements for calculating the centrifugal force due to traffic on curved bridges, where the specified force varied with the inverse of the radius of curvature.

3.6 Interim design and workmanship rules—Part 1: Loading and general design requirements, 1973 [8]

In 1973 the Merrison Committee, which had been looking into the design and erection of steel box girder bridges, published an appendix containing its recommendations for the design of such structures. Part 1 of the appendix, *Loading and general design requirements*, gave the general design principles as well as the traffic loading requirements. Steel box girders were to be designed to limit state principles against collapse and unserviceability, using partial factors on the applied loads and on material strengths. These rules were accepted by the Department of the Environment and other public bridge owners and were used as an interim measure pending the publication of the proposed new comprehensive bridge code.

The characteristic values of the various types of loading such as live load and dead load were specified as those of the appropriate design loads given in BS 153: Part 3A: 1972. One third HA was retained for application on those lanes not loaded with full HA loading. A comprehensive table of the different load combinations that were to be considered was given, together with the values of the load and material factors that were to be applied to the characteristic values. For instance, in the primary traffic loading combination the partial load factor to be applied to type HA loading was 1.5 for the limit state of collapse and 1.2 for serviceability; for type HB loading the corresponding partial factors were 1.2 and 0.9 respectively.

3.7 Department of Transport Technical Memorandum (bridges) BE 1/77, 1977 [9]

Technical Memorandum BE 1/77 was issued by the DoT in 1977 as an interim measure pending the publication of the comprehensive bridge design code being prepared under the auspices of the British Standards Institution. It was to be used in conjunction with BS 153: Part 3A: 1972 and was a revision of Technical Memorandum BE 5/73, which it superseded. As far as the requirements for traffic loading were concerned there were very few changes from BE 5/73. Highway structures, which were to be subjected to the standard highway loading, were defined for the first time as those having a span or internal diameter greater than 0.9 m. Footbridge and gantry supports were required to be able to resist a vehicle impact of 50 kN, while the superstructures of road overbridges and footbridges with a headroom clearance less than 5.5 m were required to resist an impact of 50 kN in a direction anywhere between the horizontal and the vertical. An Amendment No. 1 was issued in 1979 which clarified the application of the KEL and also the number of units of HB loading to be applied to the different classes of road.

References

1. British Standards Institution (1954) BS 153: Part 3A: 1954, *Girder bridges—Part 3: Loads and stresses—Section A: Loads*. London: BSI.
2. Henderson, W. (1954) 'British highway bridge loading', *Proceedings of the Institution of Civil Engineers*, 2 March, 325–373.

3. Statutory Instrument 1952 No. 2173 (1952) *Motor vehicles (authorisation of special types) General Order 1952*. London: HMSO.
4. Ministry of Transport (1961) Memorandum No. 771, *Standard highway loadings (amendment to Memorandum No. 577)*. London: HMSO.
5. Ministry of Transport (1970) Interim Memorandum (bridges) IM 10, *Standard highway loading*. London: MoT.
6. British Standards Institution (1972) BS 153: Part 3A: 1972, *Specification for steel girder bridges—Part 3A: Loads*. London: BSI.
7. Department of the Environment (1973) Technical Memorandum (bridges) BE 5/73, *Standard highway loadings*. London: DoE.
8. Merrison Committee (1973) *Inquiry into the basis of design and method of erection of steel box girder bridges. Report of the committee—Appendix 1, Interim design and workmanship rules—Part 1: Loading and general design requirements*. London: HMSO.
9. Department of Transport (1977) Technical Memorandum (Bridges) BE 1/77, *Standard highway loadings*. London: DTp.

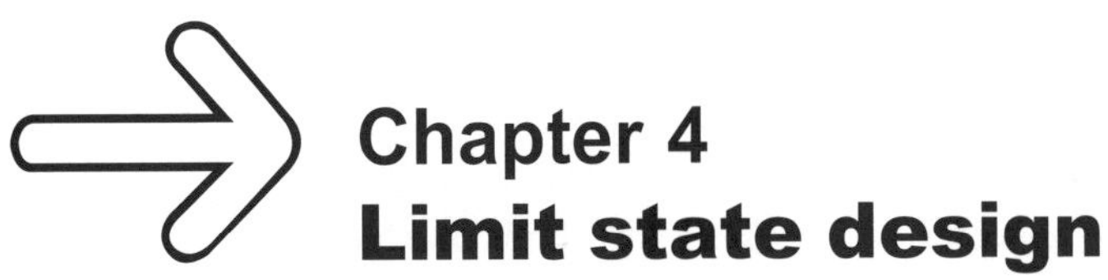

Chapter 4
Limit state design

4.1 BS 5400: Part 2: 1978 [1]

In the late 1960s the BSI decided that there was a need for a comprehensive code of practice to cover the design and specification for all types of bridges built in steel, concrete and of composite steel/concrete construction. It was also decided that the code should adopt the limit state philosophy with the use of partial factors. The code was eventually produced in ten parts under the reference BS 5400. Part 2, which covered the loading requirements for highway and railway bridges, was published in 1978. The general principles underlying the adoption of limit state design were given in Part 1. The background to the development of Part 2 was given in a paper by Henderson *et al.* to the International conference on steel box girders in London in 1973 [2].

4.2 Limit state design

Unlike the codes and standards, such as the previous loading codes, which were based on the permissible stress approach and which have in essence a single global factor of safety, the use of partial factors allows the safety margin for a particular variable to be tailored to the uncertainty associated with the evaluation of the variable concerned. Thus loads can be sub-divided into the different load types (e.g. dead, live, wind etc.) each with its own specific load factor reflecting the confidence with which that particular load can be evaluated. So dead load generally attracts a lower partial factor than live traffic loads, since the self-weight of a bridge can be calculated with more certainty than the maximum traffic loading which may be experienced by a bridge during its 120-year design life.

Limit state design requires the designer to consider the behaviour of a structure at a number of limiting states; in the case of BS 5400 the main limit states adopted were the ultimate limit state (ULS) of collapse and the serviceability limit state (SLS). The latter was generally considered to be representative of the state of the structure under normal everyday working conditions. A third limit state affecting bridges was fatigue which was covered in BS 5400: Part 10 with its own special loading requirements. The requirements for design at the ULS are usually concerned with the collapse or failure of individual elements within a structure. The seviceability limit state is less easy to define and has generally been associated with the durability of a structure. For instance in reinforced concrete design there are limitations on the size of the cracks in the concrete since wide cracks could allow the ingress of pollutants which could cause corrosion of the reinforcement and hence affect the long-term durability of the structure. The levels of loading required for the SLS are obviously lower than those required for the ULS as they are concerned with traffic situations which can occur more frequently than the very rare situation which has to be allowed for at the ultimate. While the two limit states deal with entirely different situations, there is some correlation between them in that meeting the requirements of the ULS may also satisfy the requirements of the serviceability state.

A feature of limit state design is that the value of the particular design variable for the different limit states is obtained by multiplying or dividing some basic value of the variable by the appropriate partial factor. Since statistical methods are often linked to limit state design the basic value of the variable is usually specified as its 'characteristic value', usually taken as the 5 per cent characteristic value. Thus, for traffic loading the characteristic value would be the value of the loading which has a 5 per cent chance of being exceeded during a specified period, say 100 years. A rigorous evaluation of the characteristic value of the loading for different spans would require the collection of a considerable amount of data including information on traffic flows, the frequency of traffic jams, the composition of the traffic flow and the distribution of the weights of different vehicles. These data were not available at the time that Part 2 was being developed and so the statistical approach could not be adopted; instead the basic loading was derived by deterministic methods and referred to as the 'nominal loading' rather than the 'characteristic loading'. Similarly the values of the partial factors for loads should, in a rigorous approach, have been determined by statistical methods using probability theory with the aim of achieving acceptable probabilities for the different limit states being reached. Again at the time it was not possible to do this, and therefore the values of the partial factors were determined using engineering judgement which took account of the intention that the ultimate limit state was to be a much rarer event than the serviceability limit state.

4.3 HA loading curve

The presentation of Type HA loading in BS 5400: Part 2 was similar in format to that in BS 153, consisting of a UDL varying with the span and an invariant KEL. These were to be applied in conjunction with each other to notional traffic lanes. The derivation of the formula loading followed the same procedures as that undertaken for BS 153 using trains of vehicles and determining the equivalent UDL and KEL which gave the same bending moments and shear forces. Convoys of closely spaced 24-tonne lorries were used for loaded lengths up to 30 m. For spans over 30 m the vehicle spacings were progressively increased and the 24-tonne lorries were interspersed with lighter 10- and 5-tonne vehicles. Although at that time considerably heavier vehicles

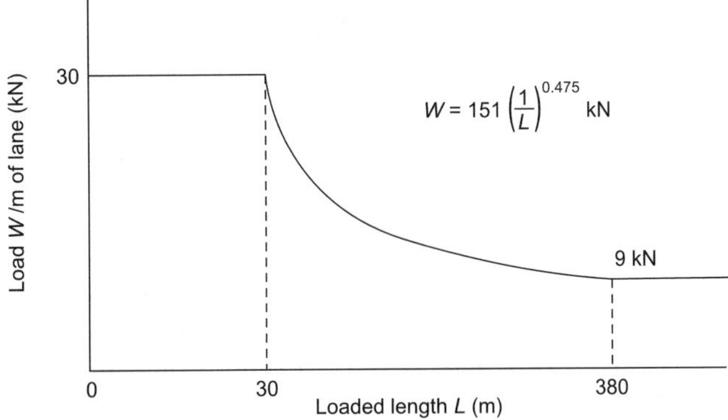

Figure 4.1. Loading curve for HA UDL [1]

were permitted, their axle weights and spacings were controlled so that their loading effects were no worse than those produced by 24-tonne vehicles. The effects of the 24-tonne vehicles themselves were no more severe than the effects of the 22-ton vehicles used in the BS 153 derivations. A 25 per cent impact allowance was added to the weight of one axle or pair of adjacent wheels in the trains, the position being chosen to give the worst load effect. The resulting uniformly distributed loading curve, which was to be used in conjunction with a 120 kN KEL, is shown in Fig. 4.1 and is very similar to that in BS 153; for ease of computation the curve was fitted to the equation $W = 151(1/L)^{0.475}$ kN where W equals the load to be applied per metre of lane and L equals the loaded length in metres.

The main difference between BS 153 and BS 5400: Part 2 was at the shorter span end where there was a horizontal cut-off at 30 m with the loading remaining constant at 30 kN/m for loaded lengths from 1 to 30 m. The effects of heavy vehicles and individual axle loads on the shorter spans were deemed to be covered by the requirement that the loading from a single 25-unit HB vehicle had also to be considered. At the long span end of the loading curve it was felt prudent to increase the minimum value from 5.8 kN/m in BS 153 to 9.0 kN/m to allow for the fact that the adoption of lower partial factors for dead compared to live load meant that any underestimate in the live loading would no longer be covered by the safety margins inherent in applying a single safety factor to both dead and live loads. Because of advances in methods of analysis, and in particular the widespread use of computers, the rather arbitrary rules for dealing with the edge stiffening of slabs were no longer thought to be necessary and were dropped.

4.4 HB vehicle

The Type HB loading vehicle for abnormal loads was retained, but instead of the distance between the innermost axles being fixed it was allowed to vary between 6 m and 25 m. This was done in recognition of the much longer abnormal vehicles which were then in service and which could produce much greater loading effects on continuous bridges than would have been generated by the short fixed length vehicle. In addition to the HB vehicle wheel loads, a single 100 kN wheel load was introduced as an alternative to the HA loading. This was intended to be used for the design of members and components with very short loaded lengths such as manhole covers and expansion joints.

4.5 Application of HA and HB loading

The concept of notional lanes for the application of the HA and HB loading was retained and expanded to cover up to six traffic lanes. Full HA loading was to be applied to two notional lanes with $\frac{1}{3}$ HA applied to the rest. The application of HB loading changed slightly from the BS 153 requirements in that the HB vehicle was no longer assumed to have sole occupancy of a lane but HA loading could also be applied to the lane to within 25 m from the front of the HB vehicle and from 25 m behind the vehicle. There was no change in the maximum weight of the HB vehicle although heavier vehicles were in use and permitted to travel under licence, since they were multi-axle vehicles whose axle weights and spacings were controlled so that their loading effects were no worse than those from the 180-tonne HB vehicle.

4.6 Partial load factors

As mentioned earlier, the HA and HB loading criteria given in BS 5400: Part 2 were described as 'nominal' loads rather than 'characteristic' loads. A rather similar simplification was made concerning presentation of the live load partial factor. This can be defined as the product of two factors, one of which takes account of the possibility of loads exceeding their nominal or characteristic values, and the other which takes account of the reduced possibility that various loads acting together would have their greatest deviations from the nominal values at the same time. In preparing BS 5400: Part 2 these two factors were not quantified separately and the partial load factor was considered as an entity for each load and load combination.

The values of the partial factors chosen for traffic loading were arrived at by engineering judgement and experience together with an evaluation of the likelihood of the various events occurring. The factor of 1.2 on HA loading for serviceability was felt to represent the difference in confidence in being able to predict the 120-year return period traffic loading compared to the confidence in predicting other loads, such as wind, for which much more statistical information was available. The factor on HA loading of 1.5 at the ultimate was arrived at by considering the possibility that three lanes on a bridge could be fully laden compared to the two lanes required by the code. On longer spans there was also the possibility that vehicles might be more closely spaced than assumed in the derivation of the formula loading, while in the shorter spans, with fewer vehicles involved, any overloading which had not been taken into account in deriving the loading would be more significant. The smaller factors on HB loading, namely 1.3 and 1.1 at the ultimate and serviceability respectively, were justified because of the controls and regulations regarding the loading and movement of such vehicles. Design studies were undertaken to compare the effects of the use of the partial load factors with the use of the implied global safety factor used in the permissible stress approach in BS 153. This was to make sure that for a member with a dead/live load ratio in the range 0.5 to 1.0 the use of BS 5400 would lead to similar designs to those generated using BS 153. The thought was that, by and large, designs done using the permissible stress approach had performed satisfactorily and appeared to have adequate margins of safety and that they could be used as benchmarks in calibrating the new code.

4.7 Load combinations

In some of the earlier codes only two combinations of the different types of loading had been specified, namely dead load combined with live traffic loading and a combination of all the different types of loads possible including wind and temperature. In BS 5400: Part 2 the second combination was split up by separating the mutually exclusive types of load and other combinations, which included secondary loads, to make five combinations in total. The five combinations were:

Combination 1 Permanent (dead) loads, earth pressure and the vertical effects of live traffic.
Combination 2 Combination 1 plus wind loading.
Combination 3 Combination 2 plus temperature loading.
Combination 4 Permanent loads plus secondary live loads together with the appropriate primary live loads associated with them. The secondary loads to be considered separately; these included centrifugal loads, braking, accidental skidding, and vehicle collisions with bridge parapets and supports.
Combination 5 Permanent loads plus those due to friction at bearings.

Details of the combinations and examples of the values of the various partial factors associated with them for selected cases are given in Table 4.1 for both the ULS and the SLS. In general the factors on primary vehicle loading were greater when these loads were considered on their own than when they were in combination with some other form of loading, such as wind, or with a secondary loading. This was to recognise that it was unlikely that the maximum values of transient loads, such as those due to traffic, wind and temperature, would occur simultaneously. The higher load factors for the ULS compared to serviceability were to ensure a bigger margin of safety against collapse than against attaining a particular state which, while undesirable, would not threaten the integrity of the structure in the short term. The value attributed to each partial factor not only reflected the level of safety or security being sought, but also reflected the confidence which could be placed in the values ascribed to a particular variable.

4.8 Centrifugal force

The centrifugal forces developed by traffic on curved in plan bridges were to be evaluated by applying point loads acting in a radial direction to the road surface. The nominal values of these point loads were to be calculated from a formula in which the main variable was the radius of curvature of the bridge. This formula had been derived from tests carried out at the Transport and Road Research

Table 4.1. Load combinations, and associated partial factors, involving traffic loads [1]

Clause number	Load	Limit state	γ_{fL} to be considered in combination				
			1	2	3	4	5
6.2	Highway bridges live loading: HA alone	ULS	1.50	1.25	1.25		
		SLS	1.20	1.00	1.00		
6.3	HA with HB or HB alone	ULS	1.30	1.10	1.10		
		SLS	1.10	1.00	1.00		
6.5	Centrifugal load and associated primary live load	ULS*				1.50	
		SLS*				1.00	
6.6	Longitudinal load: HA and associated primary live load	ULS*				1.25	
		SLS*				1.00	
	HB and associated primary live load	ULS*				1.10	
		SLS*				1.00	
6.7	Accidental skidding load and associated primary load	ULS*				1.25	
		SLS*				1.00	
6.8	Vehicle collision load with bridge parapets and associated primary live load	ULS*				1.25	
		SLS*				1.00	
6.9	Vehicle collision load with bridge supports†	ULS*				1.25	
		SLS*				1.00	

* Each secondary live load shall be considered separately together with the other combination 4 loads as appropriate.
† This is the only secondary live load to be considered for foot/cycle track bridges.

Laboratory (TRRL) by Rutley [3] in 1970 using a car and a heavy lorry. Each vehicle was driven in circles at increasing speeds until it was judged that the safe limit had been reached. At low speeds this depended upon the amount of steering lock available and at higher speeds upon the comfort of the driver and passenger. A relationship between the maximum acceptable sideways acceleration and the speed of the vehicle was found and this was used to develop the equation for the centrifugal force in terms of radius of curvature rather than vehicle speed.

4.9 Braking loads

Although the longitudinal loads specified in BS 5400: Part 2 cover both braking and traction, the forces generated by the former are the greater and therefore govern the design requirements. These were derived from a study by Burt [4] in 1969 which took account of the axle and vehicle weights of the vehicles permitted at the time under the *Construction and use regulations*, as well as the minimum required deceleration, which at the maximum vehicle weight was 0.5 *g*. It was noted however that some lorries could achieve a deceleration of 0.6 or 0.7 *g* while on dry roads the tyre to surface adhesion permitted values approaching 1.0 *g*. Braking forces for both 0.5 and 0.75 *g* were calculated for individual axles, single vehicles and short and long lengths of traffic, with progressive braking being taken into account in the last case. For the longer spans containing a long train of vehicles it was considered to be unduly severe to assume that all the vehicles would be braking simultaneously at a deceleration of 0.5 *g* and so values of 0.15 and 0.2 *g* were taken. The values of the braking forces obtained for different spans are shown in Fig. 4.2 together with the requirements for design under HA loading which were incorporated in BS 5400: Part 2. These specified a force of 8 kN/m of loaded length plus 200 kN with a

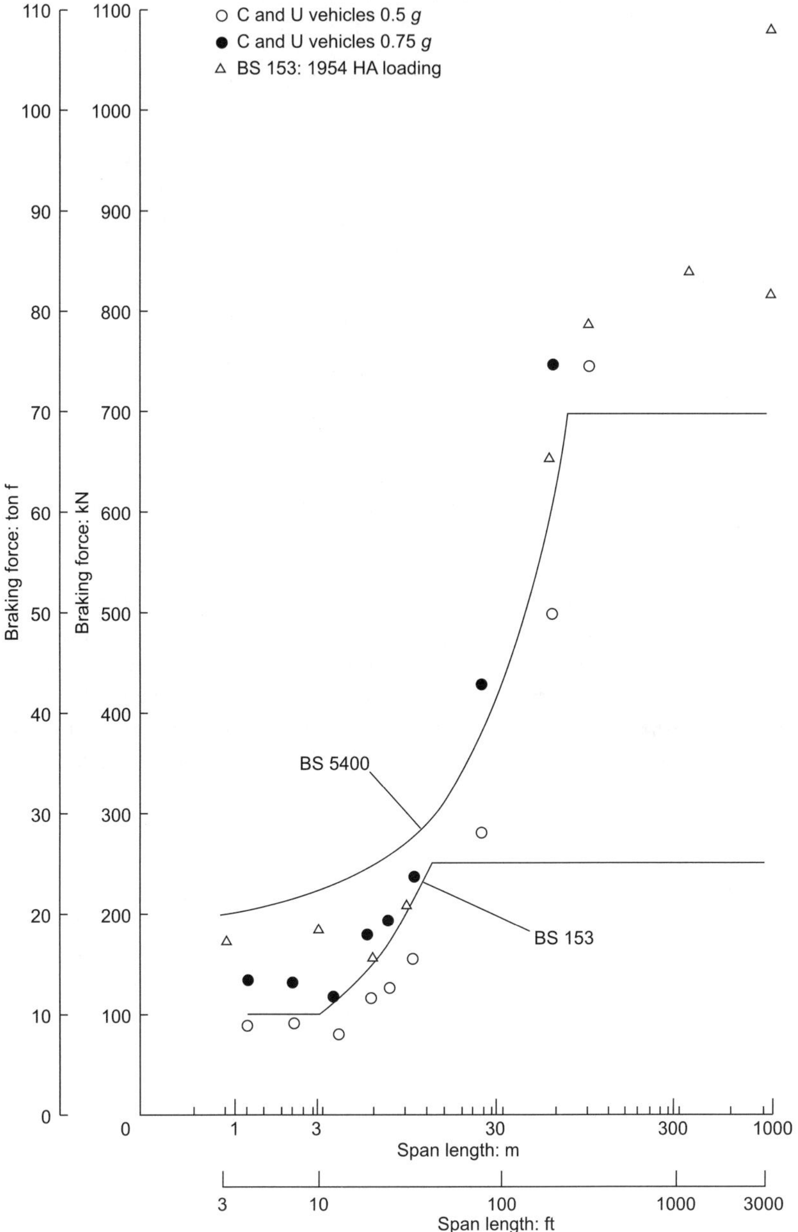

Figure 4.2. Derivation of braking loads [4]

maximum force of 700 kN per notional lane. Figure 4.2 shows that the HA design requirements adequately covered the shorter spans for axles and single vehicles with 0.75 *g* deceleration; however for the longer spans it was not considered necessary for the design rules to cover such a high deceleration bearing in mind the uncertainties over the actual mean deceleration of a convoy of vehicles. The cut-off at 700 kN was based on the assumption that for loaded lengths of about 60 to 65 m the first vehicle in a convoy of braking vehicles would have come to rest before the last vehicle had started braking, and thus there was an upper limit to the force which would be generated by a convoy of vehicles, no matter how long the convoy. This conclusion, that there was an upper limit to the braking forces on the longer spans and that the shorter spans were more than adequately covered, led to the judgement that partial load factors for the longitudinal forces of 1.25 and 1.0 for the ultimate state and the serviceability state respectively would be appropriate.

As far as the braking forces due to HB vehicles were concerned it was recognised that they were slow-moving vehicles and, because the trailers had tyres which were not designed for traction, there would be less friction developed at the road surface. Hence a lower value of deceleration, namely 0.25 *g* was selected for the HB vehicle as an entity; this assumption produced a braking force of 450 kN for 45 units of HB loading, the same as in BS 153. It was assumed that this braking force would only be applied to the two axles of one bogie to allow for the possibility of brake failure or lack of adhesion with the road surface affecting the other bogie. This was in fact equivalent to applying a deceleration of 0.5 *g* to one bogie. The low probability of a bogie failure and the fact that almost all HB vehicles had pneumatic tyres and had more than four axles, and thus could spread the braking force over a greater area, led to the adoption of load factors of 1.10 and 1.0 for the ultimate and serviceability states respectively. For both HA and HB loadings the braking forces were to be applied to one lane only, which was a judgement that while emergency braking following an incident might affect the vehicles in several lanes, the maximum braking force from one lane of closely spaced heavy vehicles travelling at speed would adequately cover the braking forces from other less heavily laden lanes.

4.10 Skidding loads

The accidental load due to skidding was derived from considering only a single errant vehicle with the skidding force being concentrated on one or two wheels. Thus BS 5400: Part 2 required a nominal horizontal force of 250 kN to be applied in any direction, the force being derived from a 250 kN vehicle decelerating at 1.0 g. The ultimate load factor was taken as 1.25 and it was judged that this, combined with the nominal skidding load, would be adequate for any unforeseen situations arising from a skidding incident; the serviceability load factor was chosen to be 1.0.

4.11 Collision loads

Loads arising from a vehicle collision with a parapet were to deal with the load that might be transmitted to the structural element to which the parapet was attached; the rules for the design of the parapets themselves were set out in other Departmental

Table 4.2. Collision loads on supports of bridges over highways [1]

	Load normal to the carriagway below (kN)	Load parallel to the carriageway below (kN)	Point of application on bridge support
Load transmitted from guard rail	150	50	Any one bracket attachment point or, for free-standing fences, any one point 0.75 m above carriageway level
Residual load above guard rail	100	100	At the most severe point between 1 m and 3 m above carriageway level

documents. The BS 5400: Part 2 rules were designed to ensure that whatever happened in an accident to the parapet or its connections to the parent structure, the structure itself would remain intact and secure. This intention was to be achieved by taking the greater of the collapse strengths of the parapet or of the connection to the parent structure as the nominal load to be applied to the supporting element. There was an associated vertical load from any four wheels of a 25-unit HB vehicle; the 250 kN vertical load was taken as representative of an errant heavy goods vehicle and including any impact effects arising from the vehicle jumping the kerb. The ultimate load factor of 1.25, and serviceability factor of 1.0, were felt to be adequate in view of there being no record of any failure of an element of a bridge as a result of a vehicle collision with a parapet.

The loads for the design of bridge supports against collisions were derived from rules developed primarily for the design of safety fences and guard rails. Two conditions were to be considered, the one where an errant vehicle struck the side of a support and the other where it struck the facing end of a support head on. These were to be considered as two separate events. In both cases two forces were to be considered, one representing the force transmitted through the supports of a guard-rail and the other representing a higher part of the vehicle coming directly in contact with the bridge support. The nominal values of these loads are given in Table 4.2 for loads normal and loads parallel to the carriageway. No primary live loads were to be applied and the load factors of 1.5 and 1.0 for ultimate and serviceability states respectively were based on judgement. It should be noted that there was a requirement that bridge supports on roads where traffic was permitted to travel faster than 80 km/h were required to be protected by a safety fence. These would extend for some distance before the support and thus help to deflect any errant vehicle away from the support, which could explain why the load parallel to the carriageway was much less than the load normal to the carriageway. From Appendix B in BS 5400: Part 2 it appears that the arrangements for the protection of bridge supports were more concerned with possible collisions from cars than from heavy vehicles.

4.12 Department of Transport, *Departmental standard BD 14/82*, 1982 [5]

The use of BS 5400: Part 2 on Department of Transport schemes was implemented by the issue of BD 14/82 in 1982. This superseded the traffic loading requirements given in *Technical Memorandum (bridges)* BE 1/77. As a result of the on-going work on the appraisal of steel box girder bridges following the Merrison inquiry [6] applicability of the Type HA loading for the longer span bridges was under review. As an interim measure BD 14/82 therefore contained the requirement that for all highway bridges where the loaded length exceeded 40 m the appropriate HA loading was to be agreed with the relevant Technical Approval Authority. This requirement was formalised by the issue of the *Interim revised loading specification* (IRLS) [7] by the Department in 1982 and which contained enhanced values of the uniformly distributed loading for longer spans.

Departmental standard BD 14/82 contained a number of amendments to BS 5400: Part 2 including the requirement that all public roads should be designed for a minimum of 30 units of HB loading, with accommodation bridges and byways being designed for a minimum of 25 units. The most significant change was in the combination of HA with HB loading where the value of the HA loading to be applied to the remaining lanes was increased from one third of the full HA to 0.6 HA. For the case where the HB vehicle occupied only one notional lane full HA was still to be applied to one other lane with 0.6 HA being applied to the rest of the lanes. The requirements for the case where the HB vehicle was taken as straddling two notional lanes caused some confusion which was clarified by the issue of Amendment No. 1 to BD 14/82 in 1983. If the width of the HB vehicle exceeded the width of a notional lane by only a small margin (i.e. one full lane and only a narrow width of the other lane was occupied by the HB vehicle) then full HA could be applied to the lane with the slight incursion. Where the HB vehicle almost fully occupied two lanes then no HA was to be applied to those lanes for the prescribed distances in front of and behind the vehicle. In both cases full HA loading could be applied to one other lane.

The design loads for vehicle collisions with parapets and bridge supports were enhanced by increasing the ultimate factor from 1.25 to 1.5. The effects of vehicle collisions with supports had now to be considered on all supports within 4.5 m of the carriageway edge irrespective of the traffic speeds. In addition to the supports the superstructures of highway and foot/cycle bridges with a headroom clearance of less than 5.7 m were to be designed for a nominal load of 50 kN.

4.13 Department of Transport's *Interim revised loading specification*, 1982 [7]

Early results from the work being undertaken by Flint and Neill [8] to revise the long-span loading showed that the requirements in BS 5400: Part 2: 1978 considerably underestimated the traffic loading on long-span bridges. In order to correct the situation as soon as possible, and to avoid any delay in waiting for the final report, the Department of Transport decided to issue an IRLS in 1982. This was done in the form of an internal memorandum to the Department's six regional offices in England; a revised version of the memorandum was issued in 1983.

The main purpose of the IRLS was to prescribe the use of considerably enhanced values for the HA UDL for loaded lengths in excess of 40 m. The revised curve is shown in Fig. 4.3 together with the BS 5400: Part 2 curve. It will be seen that the constant UDL of 30 kN/m was retained for loaded lengths up to 30 m. The revised

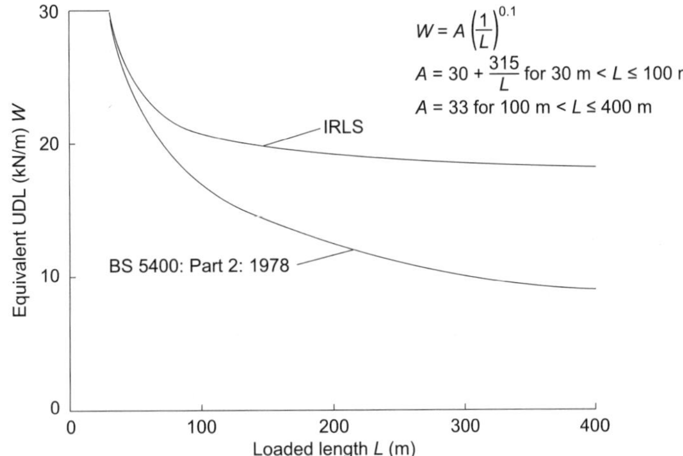

Figure 4.3. Interim revised long-span loading [7]

curve covered loaded lengths up to 400 m; for loaded lengths in excess of 400 m the loading was to be discussed with the relevant HQ division in the Department. The revised loading did not include the 10 per cent contingency margin which was to be incorporated in the later revision of BS 5400: Part 2. Although no equation was given for the loading curve, it was later determined that the following expression provided a very good fit:

$$\text{UDL (kN/m)} = A(1/L)^{0.1}$$

for spans between 30 and 400 m, where $A = 30 + 315/L$ up to 100 m, $A = 33$ for spans between 100 and 400 m and L is the loaded length.

The IRLS retained the requirements given in BD 14/82 that all further notional lanes after the first two should be loaded with 0.6 times the relevant UDL and KEL. However, there were some adjustments to the first and second lane factors which were dependent upon the loaded length concerned. For loaded lengths up to 40 m the factors were 1.0 for both lanes; for loaded lengths over 50 m the factors were 1.2 and 0.8 for the first and second lanes respectively. Between 40 and 50 m the factors were adjusted linearly in such a way that the total of the two factors equalled 2.0. There was no alteration of the HB vehicle as such but the figures in BD 14/82 showing the application of HA with HB loading were reproduced to show the values of the various lane factors for spans between 40 and 50 m and for spans over 50 m. It was noted that the figures illustrated some typical configurations of the combination of HA with HB loading but were not meant to be exhaustive.

References

1. British Standards Institution (1978) BS 5400: Part 2: 1978, *Steel, concrete and composite bridges—Part 2: Specification for loads*. London: BSI.
2. Henderson, W., Burt, M. E. and Goodearl, K. A. (1973) *Proceedings of international conference on steel box girder bridges*. Paper 11: Bridge loading. London: Institution of Civil Engineers.

3. Rutley, K. S. (1970) Transport and Road Research Laboratory Technical Note No. 455, *Drivers' and passengers' maximum tolerance to lateral acceleration*. Crowthorne: TRRL.
4. Burt, M. E. (1969) Road Research Laboratory Technical Note No. TN 401, *Forces on bridges due to braking vehicles*. Unpublished report.
5. Department of Transport (1982) Departmental Standard BD 14/82, *Loads for highway bridges—Use of BS 5400: Part 2: 1978*. London: DTp.
6. Merrison Committee (1973) *Inquiry into the basis of design and method of erection of steel box girder bridges*. Report of committee. London: HMSO.
7. Department of Transport (1982) *Interim revised loading specification (IRLS)*. London: DTp. Revised 1983.
8. Flint and Neill Partnership (1986) *Transport and Road Research Laboratory contractor report 16. Interim design standard: Long span bridge loading*. Crowthorne: TRRL.

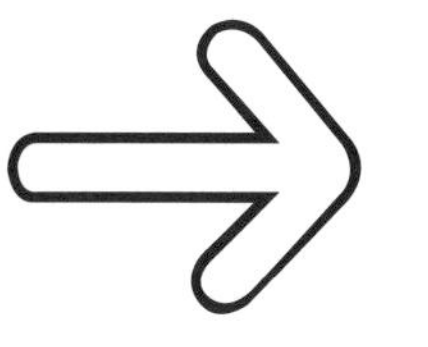

Chapter 5
Revision of HA Type loading—1. Short span

5.1 Background

The need for a review of the traffic loading on both long- and short-span bridges became apparent in the 1980s. For short spans, less than about 50 m loaded length, the catalyst was the production of a new bridge assessment code which was to be based on limit state principles with the use of partial factors. The existing assessment code [1] was based on the permissible stress approach and used a loading model derived from a limited train of idealised vehicles. It was felt that this model, which was less onerous than that used in design, did not fully reflect the loads and load effects which could be generated by the vehicles actually travelling on the highway system. In addition, for the assessment of older bridges, many of which were situated on the less well-used local roads, it was felt that it would be better if the HA loading, representing the normal Construction and Use (C&U) vehicles, was entirely divorced from the HB loading which covered the General Order and Special Order vehicles. This would mean having a HA loading curve which covered the loading requirements continuously from 1 m span upwards rather than having a cut-off at 30 m and the requirement to assess for 25 units of HB loading below that span length. A completely new loading arrangement for the shorter span bridges was therefore developed from scratch and, since it was considered that the same safety margins should apply to both design and assessment, it was decided that the same load model should apply in both cases.

The revision of the short span loading was carried out as part of the work of producing the new bridge assessment code. It was undertaken by a working party representing the major public bridge owners at the time, namely the Department of Transport, the Scottish Development Department, the Department of the Environment for Northern Ireland, the Association of County Councils, the Association of Metropolitan Authorities, British Railways Board, London Transport and the British Waterways Board. The new code, comprising a Departmental standard BD 21/84 [2] and a complementary Advice note BA 16/84 [3], was published by the Department of Transport in 1984. For design the revised short span loading was later incorporated in a revision of BS 5400: Part 2 which was published, with the full agreement and approval of the British Standards Institution, by the Department as Departmental standard BD 37/88 [4] in 1989.

5.2 Methodology

The basic assumption behind the derivation of the revised short span loading was that the worst or most extreme traffic loading that could reasonably be expected to occur in the lifetime of a bridge (i.e. 120 years) was to be taken as the extreme design loading in limit state terms, namely 1.5 × the nominal loading. Work done on the calibration of the partial factors for the steel design code BS 5400: Part 3 [5] had shown that for longer spans the 95 per cent characteristic load (i.e. that load with a 5 per cent chance of occurring in 120 years) derived from surveys of actual traffic was approximately the same as the then current serviceability loading (i.e. 1.2 × nominal HA) as specified in BS 5400: Part 2 [6]. Using the same statistical model it was shown that the ultimate design loading would occur approximately once in 200 000 years and that the nominal unfactored loading would occur once in 120 years.

Sensitivity studies were carried out as part of the partial factor calibration work to investigate the effects of changing some of the statistical assumptions regarding the live load models and other variables. These showed that the values of the resulting partial material factors obtained from the calibration procedures were relatively insensitive to the changes made to the statistical variables. Thus it was felt to be reasonable to take the ultimate HA design load as being equivalent to the most extreme rarely occurring traffic load even though it apparently had a very low probability of occurring in practice. Four components were used to generate the extreme loading effects for different spans, namely:

- (i) loading from legal C&U vehicles;
- (ii) impact;
- (iii) overloading;
- (iv) lateral bunching.

Each of these components is described in more detail below together with the details of the derivation of the HA equivalent uniformly distributed loading curve; later special reduced loads for assessment purposes were derived as described in Chapter 9. The equivalent HA loads were derived from the consideration of the loading effects for a single lane on a simply supported span; the derivation of the lane factors for the application of the loading on multi-lane carriageways is described later in Chapter 6.

5.2.1 C&U vehicle loading and impact

It was assumed that the spans would be fully occupied by convoys of vehicles laden to the limits allowed by the current *C&U regulations*. The bending moments and shear force effects generated by convoys of vehicles were calculated by a computer program, OPTAX, which selected the worst load effect i.e. Bending Moment (BM)

Table 5.1. Critical vehicles for dominant loading [2]

Span (m)	Critical vehicle* for shear force effect	Critical vehicle* for bending moment effect
10	D	C
12	D	C
14	D	C
16	B	E
18	B	E
20	B	A
22–40	B	B

Notes:
* Critical vehicles have the following characteristics:
Vehicle A: 3-axle rigid heavy goods vehicle (HGV) of 20.32 tonnes gross laden weight and 5.20 m overall length.
Vehicle B: 4-axle rigid HGV of 30.48 tonnes gross laden weight and 8.30 m overall length.
Vehicle C: 4-axle rigid HGV of 30.48 tonnes gross laden weight and 8.50 m overall length.
Vehicle D: 4-axle articulated HGV of 32.512 tonnes gross laden weight and 9.40 m overall length.
Vehicle E: 5-axle articulated HGV of 32.512 tonnes gross laden weight and 11.32 m overall length.

and Shear Force (SF), for each span length being considered. Details of the program and the particular vehicles used to make up the convoys of vehicles are given in Appendix 1. An impact factor of 1.8 was applied through OPTAX to one axle of the vehicle being considered in the single vehicle case only. OPTAX automatically applied the factor to the axle which had the most effect on the bending moment or shear force. The value of 1.8 was extracted from TRRL lab. report LR 722 [7] which gave details of impact loads measured under the rear wheels of a 2-axle rigid vehicle traversing 30 motorway bridges. A factor of 1.8 was adopted as the extreme value of impact effects obtained from the tests, ignoring a much higher value which was rejected as being a freak result. The decision to apply an impact factor only to the single vehicle case was felt to be justified because of the low probability of the impact effects of several vehicles in convoy being in phase. The various computer runs produced an envelope of the worst bending moments and shears encompassing all the then legal C&U vehicles, including the 38-tonne articulated vehicle, and including a 1.8 impact factor where appropriate. It was found that axle or bogie loading was dominant in the 1–10 m span range, multiple vehicle loading was dominant in the 25–40 m span range and there was a transition from axle to vehicle domination in between. Table 5.1 shows the particular vehicles which generated the maximum loading effects for particular spans.

5.2.2 Overloading

Information about the overloading of C&U vehicles was obtained from the results of weight surveys carried out by TRRL at three main-road sites [8, 9]. The data were collected from the static weighing of randomly selected vehicles at roadside weighbridges. A summary of the results, showing the cases of the greatest vehicle overloading, is given in Table 5.2. The greatest overloading factors were found to be 1.35 and 1.29 for a 2-axle rigid vehicle and a 4-axle articulated vehicle respectively; incidentally, these were the types of vehicle which were found to occur most

Table 5.2. Overloading—gross vehicles

Source of data		2-axle rigid	3-axle rigid	4-axle rigid	3-axle artic	4-axle artic	5-axle artic
	Assumed legal limit	16.3	24.4	30.5	24.4	32.5	32.5
Hull 1980	Max. recorded wt	18	26	33	29	42	37
	Overload factor	1.11	1.07	1.08	1.19	1.29	1.14
Litchfield 1981	Max. recorded wt	18	25	32	26	35	33
	Overload factor	1.11	1.02	1.05	1.07	1.08	1.02
Barham 1981	Max. recorded wt	22	28	31	23	36	38
	Overload factor	1.35	1.15	1.02	0.95	1.11	1.17
	Max. factor – 3 sites	1.35	1.15	1.08	1.19	1.29	1.17
Total number of vehicles surveyed		1260	185	174	178	1639	88

Note:
All weights are in tonnes.

frequently in the traffic flows. A summary of the overloading on the individual axles of the different types of vehicle is given in Table 5.3. There was some difficulty in identifying the maximum allowable weights of the various axles and bogies because these were governed by the axle spacings which were not recorded in the reported data. In the event it was decided that any axle overloading should be based on the maximum allowable under the *C&U regulations*, namely 10.2 tonnes for double-tyred wheels and 9.16 tonnes for single-tyred wheels, irrespective of the axle spacings. From Table 5.3 it will be seen that the maximum recorded axle overload factor was 1.42, which compares favourably with the 1.37 recorded for the 2-axle vehicle. Thus 1.4 was adopted as the extreme overload factor for axle weights.

Information about the weights of 4-axle vehicles was extracted from the Hull traffic data [8] and used to generate histograms of the distribution of the weights of various groups of these vehicles travelling in convoy. These histograms were processed statistically to derive the extreme loadings (i.e. mean + 4 standard deviations) due to different sized convoys of these vehicles; the extreme values obtained were then compared with the allowable legal weights of the various sized convoys to give overload factors. The resulting factors for the different sized convoys are shown in Fig. 5.1 and can be related to span length by adding up the lengths of the individual vehicles plus an allowance for headways. Although the overload factor for a single vehicle would appear to be on the low side compared to the results described earlier, the information provided by Fig. 5.1 was useful in that it indicated that overloading for a convoy of vehicles could be neglected for loaded lengths greater than about 60 m. Thus for a seven-vehicle convoy it would appear that those which were under their legal weight limit compensated for those over their weight limit. Based on all the information described above an overload factor was adopted which had a constant value of 1.4 for the axle dominated span range up to 10 m and then reduced linearly to unity at 60 m span.

5.2.3 Lateral bunching

This concept was introduced to take account of the situation where lines of traffic could be squeezed together laterally, for instance when passing a broken-down

Table 5.3. Overloading—axles

Source of data		2-axle rigid	3-axle rigid	4-axle rigid	3-axle artic	4-axle artic
	Assumed legal limit	10.2	10.2	9.15*/10.2	10.2	10.2
	2nd axle					
Hull 1980	Max. recorded wt	12.0	10.5	8.5	11.0	13.5
	Overload factor	1.18	1.03	0.93	1.08	1.32
Lichfield 1981	Max. recorded wt	12.5	9.5	6.5	13.5	14.0
	Overload factor	1.23	0.93	0.71	1.32	1.37
Barham 1981	Max. recorded wt	14.0	10.5	7.0	9.0	13.5
	Overload factor	1.37	1.03	0.77	0.89	1.32
	Max. factor – 3 sites	1.37	1.03	0.93	1.32	1.37
	3rd axle					
Hull 1980	Max. recorded wt		10.5	11.5	13.0	13.0
	Overload factor		1.03	1.13	1.27	1.27
Lichfield 1981	Max. recorded wt		10.0	11.5	9.5	14.5
	Overload factor		0.98	1.13	0.93	1.42
Barham 1981	Max. recorded wt		10.5	10.5	8.5	11.0
	Overload factor		1.03	1.03	0.83	1.05
	Max. factor – 3 sites		1.03	1.13	1.27	1.42
	4th axle					
Hull 1980	Max. recorded wt			10.5		12.5
	Overload factor			1.03		1.23
Lichfield 1981	Max. recorded wt			11.0		14.5
	Overload factor			1.08		1.42
Barham 1981	Max. recorded wt			10.0		1.05
	Overload factor			0.98		1.03
	Max. factor – 3 sites			1.08		1.42

Notes:
* 9.15 tonne limit for 2nd axle.
All weights in tonnes.

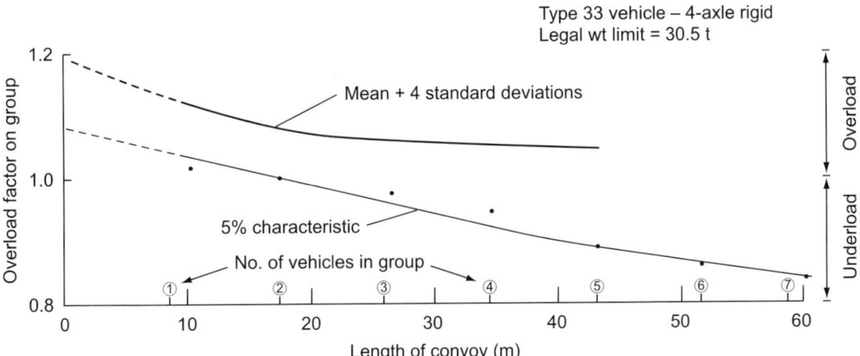

Figure 5.1. Overloading of groups of 4-axle vehicles [8]

vehicle at the side of the road. Thus three lines of vehicles could be squeezed together to occupy two notional traffic lanes and increase the loading on those two lanes. A width factor was therefore introduced which was applied to the lane loading derived from a single line of vehicles to enhance the value to be applied in design and assessment. It was based on the standard traffic lane width of 3.65 m and the maximum vehicle width allowed under the *C&U regulations* of 2.5 m. The maximum value of the factor, $3.65/2.5 = 1.46$, was intended to apply only to slow-moving jam situations and was taken as constant for spans up to 20 m. It then reduced linearly to unity at 40 m where it was considered that the chances of having a jam situation with convoys of lorries, each consisting of at least five short-wheelbase vehicles closely spaced side by side, were remote.

Because the worst width factor case was incorporated into the basic formula loading, assuming a notional lane width of 3.65 m, it was necessary to have compensating factors for the situations where the actual notional lane widths were less than 3.65 m. These correction factors, which were the reciprocal of the corresponding lane width factors, were to be applied to the relevant value of the formula loading during the actual design or assessment process.

5.3 Derivation of loading

For various spans up to 55 m the worst envelope values of bending moments and shear forces derived from the legal C&U vehicles using the OPTAX program, including the 1.8 impact factor where relevant, were multiplied by the appropriate overload and lane width factors to give the so-called extreme load effects. These load effects were then divided by 1.5 to produce the nominal effects according to the principles defined earlier in the chapter. These nominal effects were then increased by 10 per cent to allow for any unforeseen changes in vehicle design or in traffic patterns in the future. At one time it was proposed that the 10 per cent increase would be applied only for new design and not for the assessment of existing structures but, on reflection, it was considered that because many of the structures being assessed would still have relatively long service lives ahead of them it would be prudent to provide a future contingency margin for them as well. Because the 120 kN KEL was to be retained in the assessment and design codes, its effect in terms of bending moment and shear force was removed from the calculations. This left a basic nominal load effect for each span which was converted into an equivalent UDL whose value was such that, when applied over the full span length, it would give the same bending moment and shear force effects. It was found that in all cases the shear governed and the shear values were therefore used to determine the nominal formula loading for the different span lengths. It was possible to fit an equation for a curve to the individual calculated points of the form $W = 260(1/L)^{0.6}$ where W is the value of the equivalent uniformly distributed load in kN/m and L is the loaded length in metres.

The revised equivalent UDL curve is shown in Fig. 5.2 together with the curve from BS 5400: Part 2 and that from BS 153 for longitudinal beams. It should be noted that although BS 5400: Part 2 had a horizontal cut-off at 30 m, and thus apparently much lower values of loading for the shorter spans, there was a requirement to

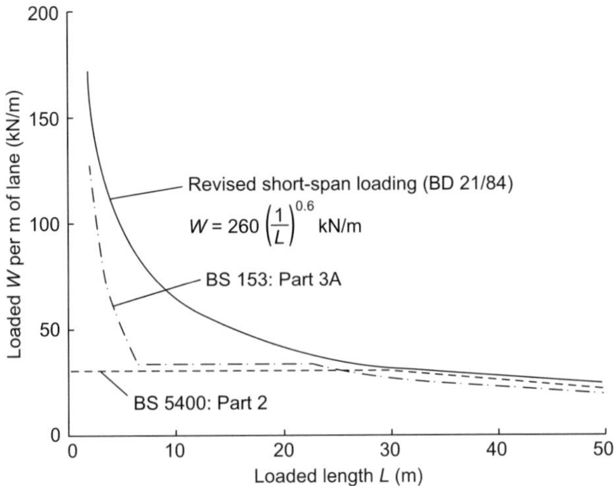

Figure 5.2. Revised HA loading curve [2]

consider 30 units of HB loading in conjunction with HA loading even for normal loading cases. The revised HA loading, which had no such requirements, was compared with 30 units of HB loading for spans up to 30 m. It was found that 30 units of HB loading adequately covered the effects derived from the revised HA loading when considering the ultimate limit state, the margin being greater when comparing shears than when comparing bending moments. The revised loading was incorporated in the new assessment code BD 21/84 [2] which was published by the DTp in 1984.

5.4 Future European vehicles

At the end of 1984 an EC Directive (85/3/EEC) [10] was issued which set out the weights and dimensions of lorries which would be able to operate on international transport anywhere within the EC. This Directive set the maximum gross weight of 5- and 6-axle vehicles at 40 tonnes; where a 40 ft ISO container was to be used the maximum gross weight was raised to 44 tonnes. In July 1986 an amending Directive (86/360/EEC) [11] set the maximum drive-axle weights for these vehicles at 11.5 tonnes. In 1989 an amending Directive 89/460/EEC [12] was issued which set out the weights and dimensions of 2-, 3- and 4-axle vehicles, all with a maximum drive axle weight of 11.5 tonnes. Details of all these vehicles are given in Table 5.4.

Work done on the effects of these proposed European vehicles, in particular the 5- and 6-axle ones, showed that they would make significant inroads into the 10 per cent contingency margin built into the revised loading curve. The loading curve, at that stage, only covered vehicles up to the UK limits of 38 tonnes gross weight and 10.5-tonne drive axles. It was therefore decided to adjust the HA design curve to cover the proposed EC vehicles and to restore the 10 per cent contingency margin. The process described above was therefore repeated with the EC vehicles included and yielded an equation for the equivalent uniformly distributed loading

Table 5.4. Summary of EC proposals on weights of certain road vehicles for international transport (February 1989)

		Tonnes
Rigid vehicles	2 axle	18
	3 axle	26
	4 axle	32
Articulated vehicles	5 and 6 axle	40*
	4 axle	38 or 36†
	3 axle (bus)	28
Drawbar trailer combinations	5 and 6 axle	40
	4 axle	36
Drive axle weight		11.5

Notes:
* 44 tonnes where semi-trailer carries a 40 ft ISO container as a combined road/rail transport equation.
† 36 tonnes with closely spaced semi-trailer axles. 38 tonnes with widely spaced semi-trailer axles.
Basic rules governing the wheelbase of 4, 5 and 6 axled vehicles: rigids and drawbars 5 tonnes/m; artics 4 tonnes/m.

of the form $W = 336(1/L)^{0.67}$. This loading, which was to be used in conjunction with the 120 kN KEL, was incorporated into the assessment code as Amendment No. 1 to BD 21/84 [2] and was issued by the DTp in 1989; it was promulgated for design purposes by the Department in Departmental standard BD 37/88 [4] in 1989. When the finalised proposals for the 2-, 3- and 4-axle vehicles were received it was found that some of the vehicles were more onerous than the vehicles which had been used to enhance the HA loading and ate into the contingency margin by up to 2 per cent for spans below 20 m and by up to 4 per cent for spans over 20 m. However, it was decided that although it was unfortunate that the long-term contingency margin had been eaten into at such an early stage the encroachment was acceptable. A critical vehicle was found to be a 32-tonne rigid vehicle with a relatively short wheelbase which, it was felt, would be unlikely to be used a great deal in international traffic to the UK.

Details of the application of the revised HA loading for short spans, including the selection of lane factors, are given in Chapter 9.

References

1. Department of the Environment (1973) Technical Memorandum (bridges) No. BE 3/73, *The assessment of highway bridges for construction and use vehicles*. London: DoE.
2. Department of Transport (1984) Departmental standard BD 21/84, *The assessment of highway bridges and structures*. London: DTp.
3. Department of Transport (1984) Advice note BA 16/84, *The assessment of highway bridges and structures*. London: DTp.
4. Department of Transport (1989) Departmental standard BD 37/88, *Loads for highway bridges*. London: DTp.

5. Flint and Neill Partnership and Imperial College (1980) *Derivation of safety factors for BS 5400: Part 3—Final report and appendices*. Unpublished report for Department of Transport.
6. British Standards Institution (1978) BS 5400: Part 2: 1978, *Steel, concrete and composite bridges—Part 2: Specification for loads*. London: BSI.
7. Page, J. (1976) Transport and Road Research Laboratory lab. report 722, *Dynamic wheel load measurements on motorway bridges*. Crowthorne: TRRL.
8. Glover, M. H. (1983) Transport and Road Research Laboratory supplementary report 770, *Results from the Hull axle weight survey (1980)*. Crowthorne: TRRL.
9. Glover, M. H. and Shane, B. A. (1983) Transport and Road Research Laboratory supplementary report 787, *Results from axle weight surveys at Lichfield and Barham*. Crowthorne: TRRL.
10. Council Directive 85/3/EEC—OJ No. L2, 3.1, 1985, p. 14.
11. Council Directive 86/360/EEC—OJ No. L217, 5.8, 1986, p. 19.
12. Council Directive 89/460/EEC—OJ No. L226, 3.8, 1989, p. 5.

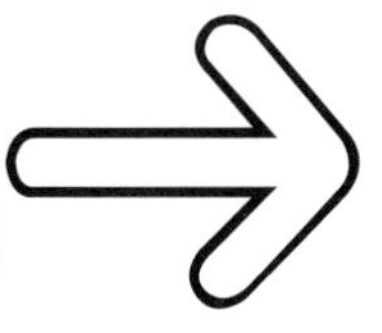

Chapter 6
Revision of Type HA loading—2. Long span

6.1 Background

The need for a review of the traffic loading on long-span bridges became apparent as a result of the appraisals of long-span steel box girder bridges, in particular the Severn Suspension Bridge, which were carried out in the wake of the Merrison Report [1]. Further work on loading was done in the development of a statistical approach for the rationalisation of the partial material factors in Part 3 of BS 5400 [2] using the techniques outlined in CIRIA Report 63 [3]. These studies showed that the Type HA loading in BS 153, which was used in the design of the Severn Bridge, and the loading requirements in the then current design code BS 5400: Part 2 [4] did not adequately cover the traffic loading actually being experienced by long-span bridges. Observations of traffic jams had shown that the spacings between vehicles were substantially less than those assumed in deriving the BS 153 and BS 5400 loading criteria. Moreover, the high proportion of the longer heavy articulated lorries in the traffic flow resulted in a lower void ratio because of the absence of the gap that there would have been between two shorter wheelbase vehicles with the same total weight. The observations also called into question the assumptions regarding the amount of dilution of heavy traffic by light vehicles such as cars and light vans as the large increase in the number of heavy goods vehicles meant that convoys containing a much greater number of heavy goods vehicles than assumed in BS 153 [5] (and BS 5400) were occurring fairly frequently. The remarkable change in freight transport by road since BS 153 was produced can be seen by reference to the statistics for the period 1962–1977. These showed the following changes during that time:

(i) goods moved by road, measured in tonnes/km, virtually doubled;
(ii) the number of goods vehicles with a gross weight over 28 tonnes rose from an insignificant number to 90 000;
(iii) the number of goods vehicles with a gross weight under 11 tonnes fell substantially.

All of these findings highlighted the fairly urgent need for the development of a more realistic Type HA design loading for longer span bridges. The necessary work was carried out for the Department of Transport by the Flint and Neill Partnership and was completed in 1983. It was later reported in TRRL contractor report 16 [6] in 1986. The recommendations were used by the Department to produce interim advice for designers in the *Interim Revised Loading Specification* (*IRLS*) [7] issued in 1983. The findings were later incorporated in a more permanent form in a revision to Part 2 of BS 5400 which was issued in 1988 by the Department as Departmental standard BD 37/88 [8].

6.2 Outline of loading study

The study involved the development of mathematical models of the traffic flows on heavily trafficked routes, such as motorways and all-purpose trunk roads, and the statistical determination of the characteristic values of the live load effects which could arise from traffic jams occurring during the 120-year design life of a bridge. The characteristic value was defined as the 5 per cent value which meant that there was a 1 in 2400 chance of the particular value being exceeded in one year. The study covered loaded lengths in the range 75–1600 m and considered both single lane and two- and three-lane carriageways. The loading models were based on traffic flows, traffic composition and vehicle weights recorded in 1980; estimated data for 1990 were produced by TRRL to take account of the growth of motorway traffic for each class of vehicle. These predictions also allowed for the introduction under the *C&U regulations* of a 38-tonne 5-axle articulated vehicle.

An important part of the study was the setting-up of a garage of vehicles whose weights and proportions by class were based on data derived from a comprehensive roadside axle weight survey and from the regular general census data obtained for the trunk road network. Vehicles were selected at random from this garage and in the preliminary studies were used to generate convoys of a specified number of vehicles which were assumed to cover a related loaded length. In later work this approach was refined to allow for lane selection by vehicles arriving at a jam, random positioning of the head of a queue or jam and various influence line shapes. Variations in loading at different times during the day because of variations in the traffic mix were also considered.

The 5 per cent characteristic values of the load effects were derived using both normal and extremal distributions and were converted into the equivalent uniformly distributed loading which was to be applied in conjunction with a KEL. The individual lane loadings for multiple-lane carriageways and for an adjacent second carriageway were investigated in order to determine the appropriate lane loading factors. The sensitivity of the results to the various assumptions were also assessed. Figure 6.1 shows, in diagrammatic form, the sources of the information used and the various processes undertaken in deriving the characteristic load effects for different loaded lengths. It will be noted that a limited amount of work was done to derive partial live and dead load factors using reliability methods and building on work done previously in the calibration studies for BS 5400: Part 3 [2].

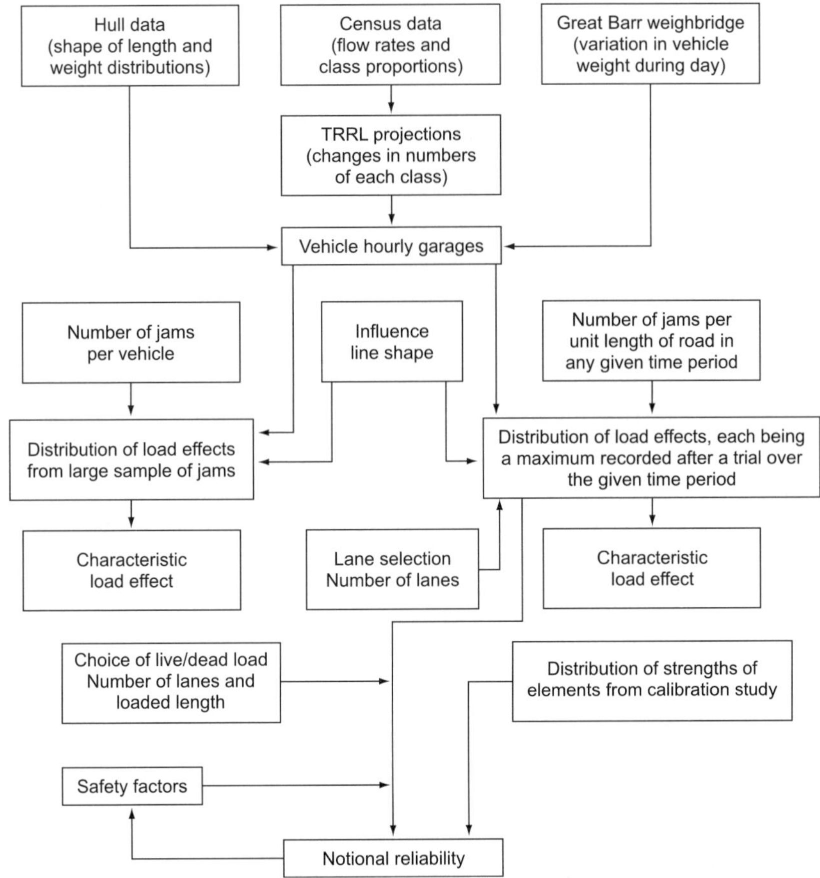

Figure 6.1. Outline of data sources and procedures used to determine characteristic load effects [Flint and Neill, presentation overhead]

6.3 Data used in study

6.3.1 Vehicle weights and lengths

The results from a static roadside survey of axle weights of lorries on the A63 at Hull in 1980 [9] were used to provide the data on the weights and lengths of all the heavy goods vehicles used in the study. This data did not include information on cars or light goods vehicles which were assumed to have constant weights and lengths of 1.2 and 3.5 tonnes/m respectively. Buses were treated as part of the 2-axle rigid group of vehicles. Each vehicle was given its own particular classification as shown in Table 6.1. Each 3- and 4-axle heavy goods vehicle type was allocated to one of two distinct sub-groups; those more than and those less than '$\frac{1}{2}$ laden' weight, the '$\frac{1}{2}$ laden' weight being defined as being midway between the allowable gross vehicle weight and a typical unladen weight of the vehicle class concerned. The percentage of laden to unladen was calculated by relating the average weight of the sub-group to the overall average weight of the particular vehicle type. The

Table 6.1. Classification of vehicle types [6]

Type No.	Type ref.	Weight (tonnes)	Length (m)	$\frac{1}{2}$ laden weight (tonnes)	GVW (tonnes)	Remarks
1	Car	1.2	4.5	–	–	
2	L.G.	3.5	5.5	–	–	Light goods
	Bus	–	–	–	–	Treated as 31, 2 R
31	2 R	H	H	–	16.2	2-axle rigid HGV
32	3 R	H	H	15.55	24.4	3-axle rigid HGV
33	4 R	H	H	19.70	30.5	4-axle rigid HGV
51	3 A	H	H	15.95	26.4	3-axle artic HGV
52	4 A	H	H	21.50	32.5	4-axle artic HGV
92	4+A	H	H	21.50	32.5	As 52, 4 A. 22 T payload
94	4+A	A	A	25.00	38.0	5-axle 26 T payload

Notes:
H: randomly selected pair of weight and length from Hull data.
A: Hull weight and length pair revised for 1982 C&U regulation vehicles.
'$\frac{1}{2}$ laden weight' is used to split vehicle types into 'laden' and 'unladen' groups to enable correct proportion of 'laden' vehicles to be selected.
31–52 are C&U vehicles to 1978 regulations.
92–94 are 'revised vehicles' being artic vehicles with 4 or more axles conforming to 1982 C&U regulations.

weights and lengths of the various Hull vehicles, adjusted where necessary, were stored in the random order in which they had been recorded at site.

6.3.2 Proportions of different vehicle types

The relative proportions of the different vehicle types were known from the regular 200-point census data to vary widely between sites and during the day. Motor vehicle travel also varied throughout the year, but travel in October was known to be close to the yearly average. A full week's traffic records for October for 11 sites from the 200-point census were selected as the basis for the traffic flows and mixes used in the study. The 11 sites were chosen as being representative of the heavy traffic conditions likely to be found on major long-span bridges, and three were selected from the 11 as being representative of the more heavily loaded bridges. These were at Lumley Dene on the A1, Kirklees on the M6, and the Severn Bridge on the M4. In some cases the averaged traffic flows and mixes from the three sites were used in the studies. From initial studies it was ascertained that 6.00–7.00 a.m. was the period which had the highest proportion of heavy goods vehicles combined with a significant flow of traffic.

6.3.3 Variation in average vehicle weights through the day

Previous studies had shown that there was a variation in the proportion of laden to unladen vehicles throughout the day. Unfortunately, the Hull data on vehicle weights did not record the times when the vehicles were weighed. However, data collected by TRRL over a three-week period on the M6 at Great Barr in 1978 gave the hourly average weights for each classified type of vehicle during the

24 hourly periods. These data were then used to calculate the percentage of 'laden' (and hence '$\frac{1}{2}$ laden') for each vehicle type and for each hour during the day and these proportions were then applied to the Hull data so that any sample of traffic generated from the Hull data had the same laden/unladen proportions. In some of the initial traffic models only 4-axle articulated vehicles were sub-divided into the laden/unladen sub-groups.

6.3.4 Spacing between vehicles

Very few data were available about the spacing of vehicles in static or slow-moving jams apart from observations of vehicles at traffic lights and queues of traffic crawling past roadworks. For the initial loading models the spacing between all the vehicles in a convoy was assumed to be constant, being 0.9 m for sets of ten vehicles or fewer and increasing linearly to 1.8 m for 100 vehicle sets. In later models the spacings were assumed to follow a truncated normal distribution between 100 and 1000 m with the truncations at 2 standard deviations from the mean. For loaded lengths less than 100 m and greater than 1000 m constant values of 1.35 m and 2.7 m respectively were assumed.

6.4 Characteristics of traffic jams

Information from studies carried out by TRRL [10] indicated that typical accident rates involving injury, and thus being serious enough to cause a subsequent jam, were of the order of 1 per million vehicle miles. On the assumption that an injury accident could delay 1000 vehicles in a slow-moving queue extending over a mile, this could mean that 0.1 per cent of the total traffic flow over a length of road or over a long-span bridge would be closely spaced. Since similar congestion could be caused by other non-recorded incidents or by non-injury accidents, it was felt that a higher proportion of the traffic flow could be involved in closely spaced jams, especially on bridges subject to high winds where tall vehicles were likely to overturn.

For the initial stages of the study fairly arbitrary values for the percentage of flows involved with jams were adopted, namely:

(i) for the period 6.00–7.00 a.m., 0.75 per cent of the flow on weekdays was assumed to be close spaced—equivalent to 1 h on two mornings each year;
(ii) for the period 11.00 a.m.–12.00 noon, 2.0 per cent of the flow on weekdays was assumed to be close spaced—equivalent to 1 h on five mornings each year.

For the later more detailed studies the following rule was adopted:

percentage of the traffic flow in close spaced jams = 2 × hourly flow/1200.

In developing the extremal loading model it was assumed that an incident causing a jam could occur at any position along the bridge concerned. Based on the arguments outlined above the rate of any vehicle causing congestion was assumed to be 1 per 50 000 miles of bridge span traversed. This led to the assumed frequency of a jam forming within any hour within a fixed length of span being 1/50 000 × length (miles) within which jam head occurs × average hourly flow rate. For the Severn Bridge this assumption implied that two jams could be expected to occur in one carriageway per year. Assuming that the jam head would form at a random position along the carriageway meant that the head of a jam could form at any point along the carriageway with the same frequency. In the absence of any recorded information it was assumed that all jams would persist for 1 h only.

6.5 Vehicle lane selection

The assumptions regarding the selection of lane by drivers approaching a jam situation were influenced by the observation that drivers tend to join the tail of the queue in the lane which contains the fewest vehicles. Also light vehicles tend to avoid lane 1 when there is a high proportion of heavy goods vehicles in the traffic. A procedure was developed to place vehicles in particular lanes which took account of the above as well as the observation that heavy goods vehicles do not normally enter lane 3 unless the queue there is very much shorter than the queues in lanes 1 and 2. For a two-lane carriageway the lane 3 conditions were omitted.

6.6 Projections of traffic growth to 1990

Because of changes to the *C&U regulations* which were to come into force in 1983, and which allowed the use of 38 tonne articulated vehicles with 5 axles, and due to the increasing volumes of traffic and increasing numbers of heavy goods vehicles (including the switch from 3- to 4-axle vehicles) it was decided to see what effect these changes might have on the traffic mix in 1990 and the consequences for bridge loading. So, in addition to developing loading models based on the 1980 Hull data, it was agreed that loading models should be developed based on projections of the likely traffic flows and proportions of the heavier vehicles in 1990. These projections were based on upper and lower bound estimates produced by TRRL of motorway flows by vehicle type. These estimates covered both the case where the maximum vehicle weight was limited to 32.5 tonnes and the case where the maximum weight had been increased to 38 tonnes, both cases being considered in the loading studies. The main changes in the composition of traffic flows forecast for 1990 compared to 1980 were the increase in the flow of rigid 4-axle vehicles and the switch from 3-axle articulated vehicles to 4-axle articulated vehicles. In modifying the projections to take account of the effects of introducing the heavier 38-tonne vehicles the following assumptions were made:

(i) the changes to the *C&U regulations* would only affect 4-axle articulated vehicles;
(ii) the total payload flow carried by 4- and 5-axle articulated vehicles would remain the same for 1990 traffic. Thus the effect of introducing 38-tonne vehicles would be to reduce the combined number of 4- and 5-axle vehicles.

In order to incorporate the above projections into the loading models it was necessary to transform the Hull garage of the then current C&U goods vehicles into a 1990 garage by revising the weights and lengths of the recorded 4-axle vehicles. The garage of Hull 1980 vehicles was scanned vehicle by vehicle with only the 4-axle articulated vehicles being selected for treatment. The selected vehicle would either remain as a 4-axle vehicle with a gross weight of 32.5 tonnes, or it would be modified to become a 5-axle vehicle with a gross weight of 38 tonnes. The choice of treatment was made randomly but in the proportions estimated by TRRL for the 1990 traffic flows. The actual weights of the modified heavier vehicles were determined by increasing the actual payload in proportion to the increase in the nominal payload and by increasing the tare weight of the vehicle itself. The recorded lengths of 4-axle articulated vehicles were also adjusted, some in accordance with changes in the *C&U regulations* which permitted increases in length from 15.0 to 15.5 m, and others which were shorter than the legal requirements were given their minimum legal lengths.

6.7 Generation of traffic streams

The next step in the process was to generate strings or garages of vehicles representative of the traffic for a particular site and for specified hours of the day. The sites used were either Lumley Dene on its own or the three-site average already mentioned. The method of assembly of these strings of vehicles is shown in Fig. 6.2. The annotation of the different garages shown in Fig. 6.2 indicates whether they were based on the 1980 recorded traffic data or on the projected data for 1990, which 'site' was used to determine traffic flow rates and proportions and the particular hour in the day to which they refer. The sequence of the different types of vehicle was chosen randomly but with the odds for each type of vehicle being chosen being in proportion to its average proportion in the traffic. The flow rates and class proportions for the various vehicles for different hours during the day were obtained from the 200-point census data. Variations in vehicle weights were determined from the Great Barr data to predict the proportions of laden/unladen vehicles at different periods throughout the day. If a car or light goods vehicle was chosen its weight and length were taken as being a fixed value as described earlier. If a heavy goods vehicle was chosen its weight and length would be taken from either the 1980 Hull data or the revised 1990 projected data. This was done by searching the Hull data until a matching vehicle of the randomly selected type, taking into account the laden/unladen distinction, was found and the weight and length of this vehicle were then stored as the next vehicle in the string or garage. The next vehicle type was then randomly selected and the search of the Hull data to find a match commenced from the last vehicle selected. In this way blocks containing 780 vehicles were generated.

Not all of the garages generated were used in the final derivation of the nominal design loading, some being used to test the reliability of the various assumptions and others to check that the sample size of vehicles would be sufficient to give a reliable estimate of the form and parameters of the distribution of the statistical load model. For instance, it was found that using the garage of data for 1980 Case A (see Figure 6.2), 771 sets of ten vehicles gave results of weight/length of sets of vehicles which approximated very closely to a normal distribution.

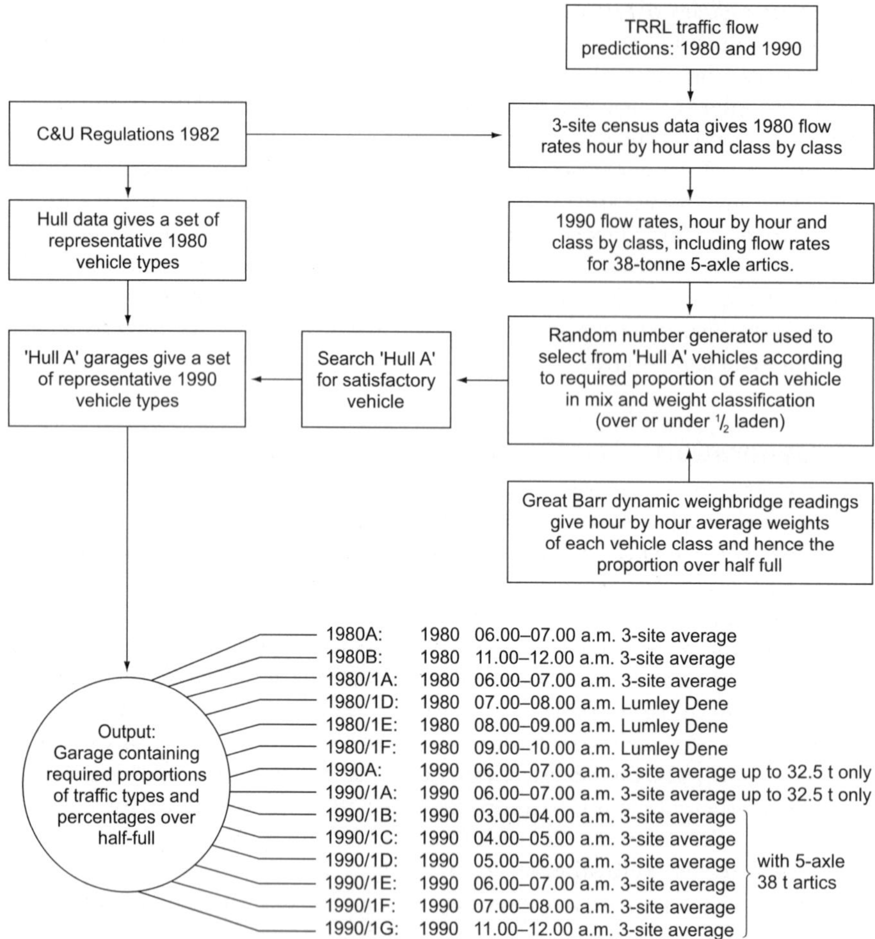

Figure 6.2. Method of assembly of traffic garages [6]

6.8 Determination of equivalent uniformly distributed load

6.8.1 Method

There were two different statistical approaches to the determination of the characteristic load effects derived from the passages of the various strings of vehicles. The first assumed that the variation of the traffic loading generated during the period 6.00–7.00 a.m. followed a normal distribution. The second assumed that the extreme loading effects generated by all the traffic during the period 3.00–8.00 a.m. followed an extremal distribution. Traffic models 1980A&B, 1980/1A to F and 1990/1A&E were used in the work assuming normal distributions while models 1990/1B to G were used for the extremal distributions. Some of the models were used in the initial work to verify the reliability of the statistical approach and to determine, for example, the number of vehicles necessary to give an accurate estimate of the parameters of the particular distribution.

6.8.2 Normal distribution

The derivation of the characteristic load effects for use in the development of the firm proposals for the design loading was done using traffic model 1990/1E. Sets of vehicles from the garage were traversed over the positive parts of a number of different shaped influence lines and the distribution of the load effects obtained for different span lengths. The lengths of the spans were determined by the number of the vehicles chosen to be in each particular set; thus a ten-vehicle set would cover a span of approximately 75 m, and a 50-vehicle set a span of approximately 400 m. In the case of model 1990/1E the definition of the frequency of occurrence of loading events was slightly more complex than used in other models and took account of significant changes in loading as the jam head moved forward. In this case the calculation to determine the number of standard deviations from the mean for the value of the characteristic load effect involved the number of vehicles in a jam and the jam length. This resulted in a characteristic value which was a slightly smaller number of standard deviations away from the mean of the distribution than obtained from the simpler calculation which regarded each jam as a loading event. The headings between vehicles were chosen randomly and assumed to vary in accordance with a truncated normal distribution and these were added to the actual lengths of the vehicles selected from the garage to model the vehicles in terms of their weight per unit length. It was noted that this approach for long spans was not suitable for short spans partly because for spans of less than about 60 m the load effects become sensitive to axle loads and axle positions. The values of the uniformly distributed loads which, in conjunction with a 144 kN KEL, were found to give load effects equivalent to the characteristic load effects in the slow lane of a 3-lane carriageway from the strings of vehicles are shown plotted for different spans in Figure 6.3.

6.8.3 Extremal distribution

This method of analysis included the effects of all traffic in the hourly periods during the period from 3.00 a.m. to 8.00 a.m. and used the maximum load effects generated by jams forming at incremental positions along a simply supported span at a randomly selected time proportional to the traffic flow in the period being considered. The traffic models, 1990/1B to G, were based on the average of the flows and class proportions from the three sites adjusted for 1990 and including the 38-tonne 5-axle articulated vehicles. The queue length caused by the jam was determined by the total flow in the period of the jam (1 h) and the arrangement and lengths of the vehicles in the garage. Load effects were calculated both for the completed stationary jam and as it began to clear with the queue moving forward by 20 per cent of the influence line length at the same spacing and composition; the maximum load effect was then selected. This process was repeated 20 times for each of the hourly periods with, in each case, the first vehicle in the jam being selected at random. For each loaded length the mode and dispersion of a Type 1 extremal distribution were fitted to the results enabling the characteristic loading effects to be calculated. The results in the form of the equivalent UDL, to be applied in conjunction with a 144 kN KEL, are shown in Fig. 6.3 for the worst lane of a 3-lane carriageway.

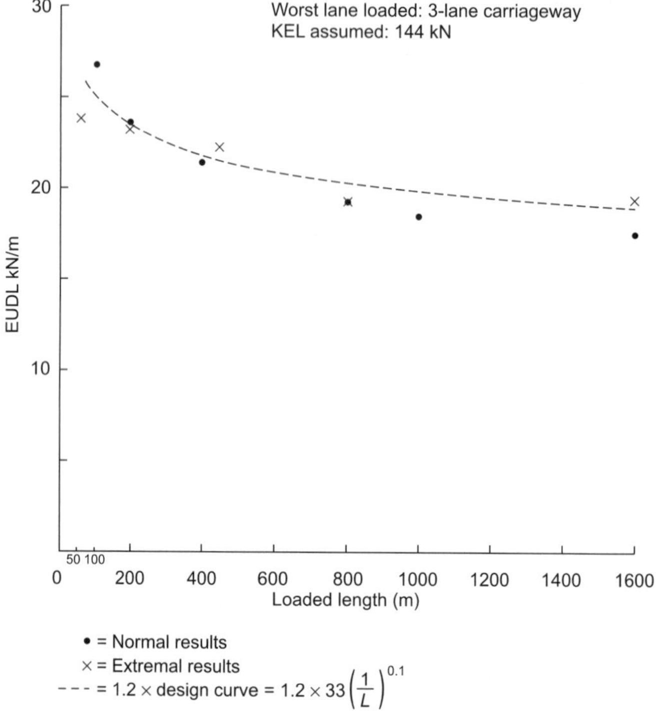

Figure 6.3. Comparison of characteristic EUDLs obtained from normal and extremal distributions

6.8.4 Lane load factors

Using the facility which allocated vehicles drawn from the garages to the various lanes in the carriageway as the jams formed, it was possible to derive the characteristic loading effects in each individual lane and hence to derive the appropriate lane factors based on lane 1 being taken as unity. These calculations were based on the 1990 traffic model adjusted for the proposed 38-tonne vehicle. The resulting lane factors for various spans and for both 3-lane and 2-lane carriageways are given in Table 6.2.

Also shown in Table 6.2 are the factors for the total load on the second carriageway of a bridge with twin carriageways. These were obtained on the assumption that the traffic flow characteristics and traffic mixes were similar in each carriageway and that the combined loading distribution for a given loaded length would have a mean value equal to the sum of the individual mean jam loads. With similar traffic conditions in each carriageway the coefficient of variation for the distribution of the combined loading was taken as $1/\sqrt{2}$ times the coefficient of variation for one carriageway. The probability of exceeding the characteristic combined loading per event was also calculated assuming a flow rate of 378 vehicles/h and that the movement of a quarter of the vehicles in a jam constituted a loading event. Based on a normal distribution the combined characteristic bending load effects for a triangular influence line were calculated for a range of loaded lengths for both twin 2-lane and

Table 6.2. Lane load factors [6]

No. of lanes	Loaded length (m)	Lane load factors			Total 2nd carriageway factor
		Lane 1	Lane 2	Lane 3	
3 lanes	100	1.0	0.73	0.49	1.44
	200	1.0	0.86	0.42	1.74
	400	1.0	0.88	0.46	1.86
	800	1.0	0.87	0.44	1.92
	1600	1.0	0.91	0.44	2.07
2 lanes	100	1.0	0.67		1.02
	200	1.0	0.67		1.14
	400	1.0	0.65		1.22
	800	1.0	0.64		1.30
	1600	1.0	0.68		1.40

Note:
Factors to be used on worst loaded lane loading

3-lane carriageways. From the combined loading effect total and knowing the loading effects for the individual lanes in the first carriageway the factors to be applied to the lane 1 loading in the first carriageway to give the total loading in the second carriageway could be calculated. These factors for various loaded lengths are given in Table 6.2.

6.9 Sensitivity studies

A number of comparative studies were carried out to test the sensitivity of the results to a number of the basic assumptions. These are described below together with the findings.

6.9.1 Percentages of heavy goods vehicles

In the study the percentages of HGVs in the traffic flows for the given hours (e.g. 6.00–7.00 a.m.) were assumed to be invariant. This assumption was tested using a normal distribution for a 3-lane carriageway carried on a 100 m simply supported span with the 1990 garage of vehicles and with 30, 60 and 100 per cent of the vehicles being HGVs. The results showed that the lane 1 and lanes 1 and 2 average mid-span moments were not significantly affected by changes in the proportion of HGVs; this was to be expected since there were sufficient numbers of HGVs in all cases to fill the lanes. Results derived for a 200 m and a 400 m 2-lane simply supported span showed that increasing the proportion of HGVs from 50 to 60 per cent increased the characteristic moments by 6.5 and 9.5 per cent respectively. For very long spans it was estimated that with 100 per cent of the vehicles being HGVs the all-lane effects would be about 25 per cent in excess of the values obtained in the study. The conclusion was that ignoring the variability of the percentage of HGVs did not result in significant errors in the derivation of the loading for bridges on heavily loaded routes carrying a high proportion of them.

6.9.2 Jam frequencies and duration

Examination of the statistical parameters derived for the extremal distributions showed that, for loaded lengths in the range between 60 m and 1600 m, a tenfold reduction in the assumed jam frequency reduced the characteristic loading by less than 5 per cent and a hundred-fold reduction by less than 10 per cent.

An analysis by the extremal method of the 2-lane average loading on a 2-lane bridge with a 1600 m long parabolic influence line showed that a reduction in the assumed jam duration of 1 h to 30 min reduced the characteristic loading by 10 per cent. For shorter loaded lengths (which could be completely filled by a jam) halving the jam duration was said to have negligible effect.

6.9.3 Traffic flow rate

For short loaded lengths, where the characteristic loading is governed by only a few vehicles in convoy, the characteristic values were said to be insensitive to flow rate. For very long loaded lengths the effects on the characteristic loading of changing the flow rates were thought to be similar to the effects brought about by changing the jam duration.

6.9.4 Vehicle spacing

Using the simple assumption for vehicle spacing of 0.9–1.8 m for 10–100 vehicle sets respectively, it was shown that for a 2-lane carriageway the characteristic average lane loading was reduced by 7 per cent for a 100 m span and by 14 per cent for an 800 m span if the spacing were doubled. For a 3-lane carriageway the all-lane average mid-span moments on simply supported spans of 100 m and 800 m were reduced by 3.5 per cent and 17.0 per cent respectively by doubling the spacing.

6.9.5 Lane distribution

The sensitivity of the lane load factors to the lane allocation rules was investigated using the normal distribution method for a 200 m simple span with 3 lanes. Allocating all HGVs to lane 1, except when the queues in lanes 2 and 3 were much shorter, resulted in insignificant change in the resulting lane factors. Allocating vehicles randomly to the three lanes resulted in a 7 per cent fall in the lane 1 loading.

6.10 Design values

The nominal Type HA equivalent uniformly distributed loading (EUDL) was based on the results of the normal and extremal methods, which are shown in Fig. 6.3. It was derived from the characteristic values shown there by dividing them by 1.2. Because there was much better agreement between the lane 1 loadings in 2- and 3-lane carriageways in comparison with the lane 2 and 2-lane average loadings it was decided that the design loading should be based on the lane 1 (slow lane) values. A good fit for the EUDL to the extremal values, which were found to slightly exceed the normal values, was given by the formula EUDL (kN/m) $= 33(1/L)^{0.1}$ where L is the length of the positive part of the influence line in metres. This design loading, which was to be used in conjunction with a 120 kN KEL, was based on the bending moment effects. As the design values calculated were based

Table 6.3. Proposed lane load factors [6]

No. of lanes	Lane load factors			Total load in 2nd carriageway
	Lane 1	Lane 2	Lane 3	
1	1.0			0.5
2	1.0	0.67		1.0
3	1.0	0.9	0.5	1.5

on predictions of traffic flows and vehicle mixes only up to 1990, it was decided that it would be prudent to increase the UDL for design by 10 per cent to allow for any future changes in traffic patterns, but keeping the KEL at 120 kN. The formula for the equivalent uniformly distributed load was therefore adjusted to be EUDL $(\mathrm{kN/m}) = 36(1/L)^{0.1}$.

The determination of suitable lane load factors by statistical means has been described and the results obtained are given in Table 6.2 for both 2-lane and 3-lane dual carriageways. These results were used to form the proposals for use in design given in Table 6.3. The proposals for the first carriageway match the calculated values fairly closely; however, there appears to have been some difficulty in matching the factors for the second carriageway, where the results in Table 6.2 would suggest that the total load factor should vary with both the loaded length and the number of lanes. The results indicate that the total design loading on the second carriageway should range from 0.6 to 0.9 times that on the first carriageway. In the event the factors proposed for design for 2- and 3-lane second carriageways represent about 0.6 times the total design loading on the first carriageway. This probably took account of the need in the longer span bridges to consider the loading conditions on the second carriageway resulting from lane closures and contra-flow working which could more than cover the apparent deficiency in the second carriageway lane factors.

The loading proposals from the long-span loading study were first issued as an interim measure by the Department of Transport as *Interim revised loading specification for bridges* [7] in 1982. They were formally adopted by the Department by the issue of Departmental standard BD 37/88 [8] in 1989, which also contained the revised short-span loading.

As part of the loading study, the effect of applying the UDL to the full length of the positive part of a highly cusped influence line was investigated. It was found that such an approach could underestimate the loading effects, which were derived from the use of one of the 1990 load models, by up to 12 per cent. A method of dealing with this anomaly by the use of a shorter base length was recommended and later incorporated in BD 37/88.

It was found during the derivation of the EUDL from the traffic loading effects that the best approximation to the maximum shears and bending moments and to the total load was obtained using a KEL which varied with the span. However, it was shown that the adoption of an invariant KEL, in conjunction with the span dependent UDL, when applied to the influence line for a simply supported span, gave a very satisfactory fit to the traffic load effects.

References

1. Merrison Committee (1973) *Inquiry into the basis of the design and method of erection of steel box girder bridges*. Report of the committee. London: HMSO.
2. Flint and Neill Partnership and Imperial College (1980) *Derivation of safety factors for BS 5400: Part 3. Final report and appendices*. Unpublished report for the Department of Transport.
3. Construction Industry Research and Information Association (1977) CIRIA Report 63, *Rationalisation of safety and serviceability factors in structural codes*. London: CIRIA.
4. British Standards Institution (1978) BS 5400: Part 2: 1978, *Steel, concrete and composite bridges—Part 2. Specification for loads*. London: BSI.
5. British Standards Institution (1954) BS 153: Part 3A: 1954, *Girder bridges—Part 3: Load and Stresses—Section A: Loads*. London: BSI.
6. Flint and Neill Partnership (1986) *Transport and Road Research Laboratory contractor report 16, Interim design standard: Long span bridge loading*. Crowthorne: TRRL.
7. Department of Transport (1982) *Interim revised loading specification (IRLS) for bridges*. London: DTp. Revised 1983.
8. Department of Transport (1989) Departmental standard BD 37/88, *Loads for highway bridges*. London: DTp.
9. Glover, M. H. (1983) Transport and Road Research Laboratory supplementary report 770, *Results from Hull axle weight survey (1980)*. Crowthorne: TRRL.
10. Johnson, H. D. (1969) Road Research Laboratory Report LR 283, *Road accidents and casualty rates based on the 1300 point census*. Crowthorne: RRL.

Chapter 7
Composite version of BS 5400: Part 2

7.1 Background

The publication of the proposed revisions to the HA loading curves for both short and long spans (see Chapters 5 and 6) prompted the British Standards Institution to initiate a review of BS 5400: Part 2: 1978. The committee responsible agreed a number of major amendments including the revision of the HA loading curve. As an interim measure and bearing in mind the considerable amount of work that was going into the preparation of a comprehensive series of *Eurocodes* for the design of structures, including bridges, it was agreed that the amendments should be issued by the Department of Transport rather than by BSI. Because of the large volume of technical and editorial changes involved, it was decided that a complete revised version of BS 5400: Part 2 should be issued rather than just a set of amendments. This so-called 'composite' version included the railway bridge loading, although this was unchanged. The composite version was issued as Appendix A to Departmental standard BD 37/88 [1] in 1989 and superseded BD 14/82 [2]. Departmental standard BD 37/88 also contained some additional requirements regarding the use of the composite version on the Department's schemes.

7.2 Type HA loading

The derivation of the revised short- and long-span loading has been described in Chapters 5 and 6 respectively, together with the equations which were fitted to the two portions of the HA UDL loading curve. Since it was considered that the method employed to model the long-span loading did not adequately model the behaviour of the traffic on short spans, it was suggested that where there was an overlap between the short- and long-span curves below about 60 m the short-span values should be adopted. In the event the two equations intersected very close to the 50 m span mark and this defined the transition point. Both equations contained a 10 per cent contingency enhancement and in both cases the UDL was to be applied

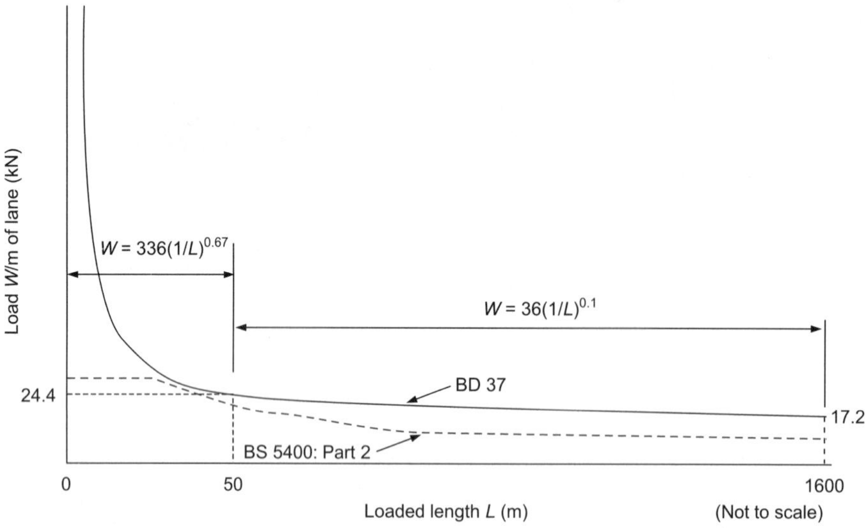

Figure 7.1. Loading curve for HA UDL [1]

in conjunction with a 120 kN KEL. The revised loading curve, which covered loaded lengths up to 1600 m, is shown in Fig. 7.1 together with the curve from the 1978 version of BS 5400: Part 2 [3]. It will be noticed that for the longer spans the value of the UDL had almost doubled. It will also be noticed that the horizontal cut-off below 30 m span had been removed since the much enhanced values of the UDL now fully covered the effects of all C&U vehicles down to a 2 m loaded length. As an alternative to the use of the UDL plus KEL the 100 kN wheel load with a contact pressure of 1.1 N/mm^2 could still be applied where appropriate.

7.3 Type HB loading

The specification of the HB vehicle was identical to that in the 1978 version with the provision for adjusting the length of the wheelbase. The amendment in BD 14/82 which clarified the lane loadings to be applied when the HB vehicle straddled two lanes was also incorporated, including the application of 0.6 HA in the remaining lanes other than those carrying the HB vehicle or full HA. However, as will be discussed, the value of the HA loading could be modified if the notional lanes were less than the standard width of 3.65 m.

7.4 HA lane factors

The lane factors for multiplying the relevant values of the HA UDL and KEL are given in Table 7.1. At first sight these appear to be rather complicated, but this is partly due to the different approaches adopted in the derivation of the short- and long-span loading. The short-span loading took account of the possibility that a

Table 7.1. HA lane factors [1]

Loaded length L (m) *	First lane factor β_1	Second lane factor β_2	Third lane factor β_3	Fourth and subsequent lane factor β_n
$0 < L \leq 20$	α_1 †	α_1	0.6	$0.6\alpha_1$
$20 < L \leq 40$	α_2	α_2	0.6	$0.6\alpha_2$
$40 < L \leq 50$	1.0	1.0	0.6	0.6
$50 < L \leq 112$ $N < 6$	1.0	$7.1/\sqrt{L}$	0.6	0.6
$50 < L \leq 112$ $N \geq 6$	1.0	1.0	0.6	0.6
$L > 112$ $N < 6$	1.0	0.67	0.6	0.6
$L > 112$ $N \geq 6$	1.0	1.0	0.6	0.6

Notes:
* N shall be used to determine which set of HA lane factors is to be applied for loaded lengths in excess of 50 m. The value of N is taken as the total number of notional lanes of the bridge (this shall include all the lanes for dual carriageway roads) except that for a bridge carrying one-way traffic only, the value of N shall be taken as twice the number of notional lanes on the bridge.
† $\alpha_1 = 0.274b_L$ and cannot exceed 1.0. $\alpha_2 = 0.0137[b_L(40 - L) + 3.65(L - 20)]$, where b_L is the notional lane width (m).

notional lane could be occupied by more than one vehicle if they were travelling closely side by side (this was known as lateral bunching). This led to the nominal HA loading being increased, as described in Chapter 5, by a factor which was constant from 0 to 20 m and then reduced linearly from 20 to 40 m. The full value of the factor was 3.65 divided by 2.5, i.e. 1.46, representing the standard lane width divided by the maximum width of an HGV. The lane factors in Table 7.1 therefore have to deal with cases where the notional lane width is less than 3.65 m, otherwise the enhancement to take account of lateral bunching will be overdone. For spans up to 20 m this is done by the application of α_1, which is just the lane width divided by 3.65. For spans between 20 and 40 m, where the width factor implicit in the loading is reducing to unity, the expression for the correcting factor α_2 is more complex but represents the lane width factor at 20 m minus the proportional decrease to the span length concerned.

The lane factors for spans beyond 50 m are those proposed in Table 6.3 from the long-span studies [4] but with some slight rounding up. The split at 112 m is determined by the choice of expression used to ensure a smooth transition in the value of the 2nd lane factor, for a 2-lane carriageway, from 1.0 at 50 m to 0.67 at about 100 m. The expression adopted, $7.1/\sqrt{L}$, has a value of 1 when $L = 50$ m and a value of 0.67 when $L = 112$ m.

It should be noted that the rules for the determination of the number of notional lanes have been slightly adjusted to be consistent with the basis for the derivation of the lane width factor. Thus notional lanes are not to be less than 2.5 m wide nor more than 3.65 m wide reflecting the maximum vehicle width under the *C&U regulations* and the standard traffic lane width on trunk roads respectively. Where a carriageway

Table 7.2. Load combinations, and associated partial factors, involving traffic loads [1]

Clause No.	Load	Limit state	γ_{fL} to be considered in combination				
			1	2	3	4	5
6.2	*Highway bridges live loading:*						
	HA alone	ULS	1.50	1.25	1.25		
		SLS	1.20	1.00	1.00		
6.3	HA with HB or HB alone	ULS	1.30	1.10	1.10		
		SLS	1.10	1.00	1.00		
6.5	Footway and cycle track loading	ULS	1.50	1.25	1.25		
		SLS	1.00	1.00	1.00		
6.6	Accidental wheel loading	ULS	1.50				
		SLS	1.20				
6.7.1	*Loads due to vehicle collision with parapets and associated primary live load:*						
	Local effects: parapet load						
	Low and normal containment	ULS*				1.50	
		SLS*				1.20	
	High containment	ULS*				1.40	
		SLS*				1.15	
	Associated primary live load:						
	Low, normal and high containment	ULS*				1.30	
		SLS*				1.10	
6.7.2	**Global effects: parapet load**						
	Massive structures:						
	Bridge superstructures and non-elastomeric bearings	ULS*				1.25	
	Bridge substructure and wing and retaining walls	ULS*				1.00	
	Elastomeric bearings	SLS*				1.00	
	Light structures:						
	Bridge superstructures and non-elastomeric bearings	ULS*				1.40	
	Bridge substructures and wing and retaining walls	ULS*				1.40	
	Elastomeric bearings	SLS*				1.00	
	Associated primary live load:						
	Massive and light structures:						
	Bridge substructures, non-elastomeric bearings, bridge superstructures and wing and retaining walls	ULS*				1.25	
	Elastomeric bearings	SLS*				1.00	
6.8	*Vehicle collision loads on bridge supports and superstructures:*						
	Effects on all elements excepting non-elastomeric bearings	ULS*				1.50	
	Effects on non-elastomeric bearings	SLS*				1.00	
6.9	Centrifugal load and associated primary live load	ULS*				1.50	
		SLS*				1.00	
6.10	*Longitudinal load:*						
	HA and associated primary live load	ULS*				1.25	
		SLS*				1.00	
	HB and associated primary live load	ULS*				1.10	
		SLS*				1.00	
6.11	Accidental skidding load and associated primary live load	ULS*				1.25	
		SLS*				1.00	

*Each secondary live load shall be considered separately together with the other combination 4 loads as appropriate.

in a narrow road can accommodate only a single notional lane this is to be taken as 2.5 m wide for the application of the UDL and KEL, with the remainder of the carriageway being loaded with 5 kN/m^2.

7.5 Load combinations and partial factors

The load combinations and associated partial factors for traffic loads are generally the same as those set out in BS 5400: Part 2: 1978 and are shown in Table 7.2. However, there is an additional form of loading under Combination 1, called 'accidental wheel loading', and the descriptions of the loads arising from collisions with parapets have been considerably expanded.

7.6 Accidental wheel loading

This loading is intended to be applied on its own to elements of a structure supporting certain areas, such as footways or cycle tracks, which are not protected from vehicular traffic by an effective barrier. The loading consists of a pair of 100 kN wheels and a pair of 75 kN wheels with the axles 1.25 m apart and with a track width of 1.8 m. The partial factors on this loading were 1.5 and 1.2 for the ULS and SLS respectively.

7.7 Loads due to vehicle collisions with parapets

These are now sub-divided into the local effects on the structural elements in the vicinity of the parapet supports, and the global effects on the bridge superstructure, bearings and sub-structures. The distinction is made in recognition of the increasing use of high containment parapets which are designed to contain HGVs. Because of their massive strength such parapets and their supports are less likely to be damaged locally, but the impact could have an effect on the structure as a whole or its supports. The global impact loads to be applied consist of a 500 kN horizontal transverse load, a 100 kN horizontal longitudinal load and a 175 kN vertical load. The sub-divisions of the various types of structure affected are shown in Table 7.2 together with the associated partial factors.

7.8 Secondary live loads

(i) Collision loads on supports: the values of the loads and their application are unchanged from BS 5400: Part 2: 1978.

(ii) Centrifugal loads: the expression for deriving the nominal centrifugal load on a curved carriageway is the same as in the 1978 version, varying inversely with the radius of curvature. However, the constant has been increased by a third presumably in recognition of the increase in the maximum weight of heavy lorries.

(iii) Braking loads: the maximum values have been increased from 700 to 750 kN.

(iv) Skidding loads: these have been increased from 250 to 300 kN.

7.9 Department of Transport implementation

Although the composite version of BS 5400: Part 2 was issued and adopted by the Department, there were a few areas where the Department had additional requirements. For instance, the loading requirements for certain specialised structures such as sign/signal gantries and buried structures were covered by their own Departmental standards. A minimum of 30 units of HB loading was specified to be applied to all trunk road bridges except for accommodation bridges and with 37.5 units to be applied on principal roads. The Department's implementation also highlighted the need to consult the relevant Technical Approval Authority in certain situations, for instance to agree the loading requirements where the superstructure was carrying more than seven traffic lanes. It also drew attention to the advisability of considering more severe collision loads on bridge supports and superstructures than those specified in BS 5400: Part 2 which provided for the most frequent type of vehicle impacts and not necessarily for those from the heaviest goods vehicles.

7.10 *Design manual for roads and bridges*, BD 37/01 [5]

Departmental standard BD 37/88 [1] was subsequently incorporated into the *Design manual for roads and bridges* (DMRB) but was superseded by the issue of a revised version, BD 37/01, in 2001. The main reason for the revision was the publication of a new load code for wind, BS 6399: Part 2 [6], which led to the need for major amendments to the relevant clauses in the composite version of Part 2. There were minimal amendments as far as traffic loading was concerned apart from updating the requirements for vehicle collision loads on supports and superstructures. Although the existing requirements were retained for use in certain circumstances the designer was referred to the appropriate standard in the DMRB, namely BD 60/94 [7] which replaced part of BD 37/88 [1]. The Department's own additional requirements were also slightly amended to cover the braking and traction loads caused by abnormal indivisible loads.

References

1. Department of Transport (1989) Departmental standard BD 37/88, *Loads for highway bridges*. London: DTp.
2. Department of Transport (1982) Departmental standard BD 14/82, *Loads for highway bridges—use of BS 5400: Part 2: 1978*. London: DTp.
3. British Standards Institution (1978). BS 5400: Part 2. *Steel, concrete and composite bridges—Part 2: Specifications for loads*. London: BSI.
4. Flint and Neill Partnership (1986) *Transport and Road Research Laboratory Contractor Report 16. Interim design standard: long span bridge loading*. Crowthorne: TRRL.
5. Highways Agency (2001) *DMRB* Vol. 1, Section 3, Part 14: BD 37/01. *Loads for highway bridges*. Norwich: HMSO.
6. British Standards Institution (1997) BS 6399: Part 2: 1997, *Code of practice for wind loads*. London: BSI.
7. Highways Agency (1994) *DMRB* Vol. 1, Section 3, Part 5: BD 60/94. *The design of highway bridges for vehicle collision loads*. Norwich: HMSO.

Chapter 8
Short-span assessment loading

8.1 Background

In 1987 the then Department of Transport launched a 15-year bridge rehabilitation programme for trunk road bridges in the UK. This was to deal with a number of different problems which had been identified over the years. Other bridge owners, such as the local authorities, were encouraged to adopt a similar programme for their bridge stock. The programme was supported by the publication of a number of documents which provided direction on the way that the programme was to be carried out as well as providing suitable technical criteria. Part of the programme was concerned with the assessment of the safe load-carrying capacity of certain types of bridge such as the older ones and those believed to have design deficiencies. The technical criteria for the assessment of these structures were originally published in Departmental Standard BD 21/84 [1] which contained the revised Type HA loading requirements for short spans whose derivation is described in Chapter 5.

The assessment of the safe load carrying capacity of existing and particularly older structures can be difficult, especially if the structure fails to meet some of the assessment criteria. This problem became more apparent as the rehabilitation programme progressed when structures, which appeared to be able to cope perfectly safely with the current traffic, were found to be sub-standard according to the assessment criteria. Many assessment engineers felt that the assessment codes, which dealt with structures in a general way, were too conservative and did not take proper account of local situations. It was recognised that whilst design codes had to have some degree of conservatism in order to cover the unknown situations in which they might be applied in the future, such a degree of conservatism was not required in the assessment of an existing structure in a known situation.

The experience gained from the rehabilitation programme led the Highways Agency to consider the need for a new approach to bridge assessment which would be more flexible than the methods then in use. It was felt that the use of probability theory might provide a way of taking account of the risks involved and the options available in dealing with sub-standard structures. One of the areas identified for investigation was the need for bridge specific live-load requirements which could take account of the local traffic and the state of the road surface on the bridge in question. Accordingly a contract was arranged with the Flint and Neill Partnership for the development of a site specific traffic load model for use in the assessment of short-span bridges. This work also involved TRL and Imperial College, London.

8.2 Outline of methodology

The procedures used in the development of the assessment live loading are described in a project report for the Highways Agency [2]. The work was done in three distinct stages. The first stage was to produce a probabilistic live load model based on a large sample of vehicle weight and length data recorded at a motorway site. These data were supplemented by measurements of the wheel forces on road surfaces from moving vehicles in order to model the dynamic effects of vehicles running on different types of road surface. The two sets of information were used to produce probabilistic load models inclusive of the static and dynamic effects. These models were then used in the determination of characteristic load effects and in later reliability analyses.

In the second stage, a set of reliability analyses were carried out on a range of pre-designed short-span bridges for six different traffic scenarios. These scenarios

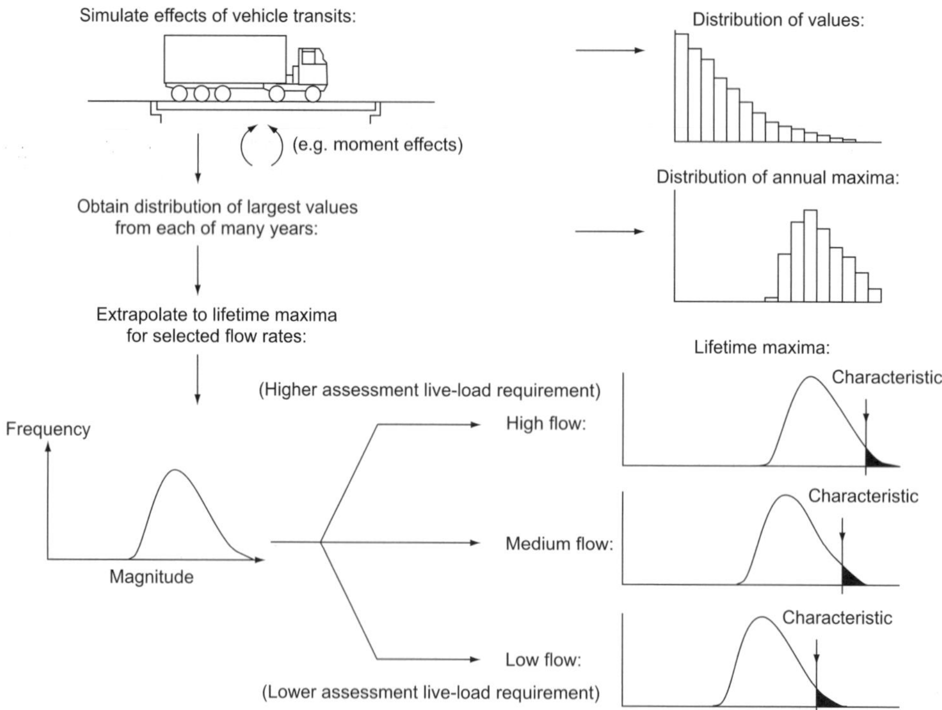

Figure 8.1. Outline of procedures for revision of short-span assessment loading [2]

covered high, medium and low traffic flows and poor and good road surfaces. From the analyses the values of reliability for the most onerous cases (i.e. high traffic flow and poor road surface) were selected and used as target values to determine suitably modified partial load factors which, when applied to the other scenarios, would give similar reliability. In the third stage, the reduced partial factors together with comparisons of characteristic load effects were used to derive load reduction factors which could be applied to the less onerous traffic scenarios. The procedures are illustrated in Fig. 8.1.

8.3 Static load models

8.3.1 Vehicle data

The static load model was based upon vehicle data collected over a two-week period in 1990 on the M6 on the northbound approach to the Thelwall Viaduct [3]. Data were obtained from a weigh-in-motion system which was installed in one carriageway of the M6 and which could record information for each vehicle passing over detectors, including speeds, axle spacings and axle weights. From the recorded data it was possible to calculate the overall lengths of vehicles, the headway between vehicles, and their axle and gross weights; it was also possible to allocate each vehicle to a particular class such as rigid 3-axle HGV or articulated HGV with a 2-axle tractor and a 3-axle semi-trailer. The full data was recorded for lanes 1 and 2 with only a vehicle count for lane 3. In all details of approximately 255 000 vehicles were recorded with the date and time of each vehicle being noted. The equipment was found to be unreliable at low speeds and was programmed not to record when speeds fell below about 35 mph. Data were therefore not available for the times when the M6 was most congested. Because there was also a video camera at the site it was possible to filter out of the data any vehicles which were considered to be abnormal or 'special types' and outside the *C&U regulations*.

8.3.2 Traffic modelling

The vehicles in the data were run in the order in which they were recorded over three different influence lines to obtain histograms of the individual and joint lane load effects. The three influence lines represented mid-span bending moments in a simply-supported beam, end shears in a simply-supported beam and bending moments over the central support of a two-span bridge beam. The traffic simulations for each influence line shape were performed for several loaded lengths from 2.5 to 50.0 m. The magnitude of each single loading event was defined as the peak load effect generated between the times that individual successive vehicles left the loaded length. Each histogram, which contained the load effects from two weeks of traffic, was then converted into a cumulative distribution function (CDF) which was then raised to the power of the daily traffic flow to give the CDF of the daily maxima. From this distribution the parameters for the best fit Extremal Type 1 distribution were obtained representing the distribution of maximum load events for two-week periods. This information was then used to derive the parameters of the Extremal distribution for any desired return period for various load effects related to particular loaded lengths, influence line shapes and traffic lanes. The characteristic load effect values (i.e. the 1 in 2400 values) were recalculated for reduced traffic flows to assess the reductions in loading for lightly used routes.

8.4 Dynamic load model

8.4.1 Data

In order to provide the necessary data for the development of a statistical model of the dynamic amplification of load effects from moving vehicles on different road surfaces, TRL carried out a series of live tests on a number of bridges using specially instrumented vehicles [3]. Four articulated vehicles and one rigid vehicle were used in the tests in which the dynamic wheel/road surface loads on each wheel were continuously recorded. The vehicles were driven over 28 selected bridges, over three stretches of road with good, medium and bad surface profiles and over three planks with different profiles placed on a smooth road surface. The bridge sites were chosen to be representative of typical motorway bridges, motorway underbridges and non-motorway bridges. In addition to the wheel load data the road surface profiles at the 28 bridges and the three test lengths of good, medium and poor surface quality were measured using high-speed surface recording vehicles.

Two of the 28 bridges were then selected and instrumented in order to measure their dynamic responses to the test vehicles. Displacements were measured at both bridges and at one of them the strains in a beam were also measured. The vehicle instrumentation was logged simultaneously with the measurements from the bridges themselves as the test vehicles traversed the two bridges at 10 and 40 mph.

8.4.2 Dynamic effect model

The wheel load data from the tests on the 28 bridges were used in the dynamic response analysis of a number of simple bridge models which were assumed to represent a range of different types of construction. These analyses were then used to prepare a histogram of all the dynamic amplification factors (DAFs) calculated for the vehicle wheel forces as recorded. The DAF was defined as the ratio of the largest load effect (including the dynamic response) to occur in a bridge transit to the largest static effect in that transit. A normal distribution curve was fitted to the histogram and was found to have a mean of 1.08 and a standard deviation of 0.15.

In order to confirm the adequacy of the theoretical calculations some of the vehicles used in the 28 bridge tests were run over the two specially instrumented bridges. Both of these were three-span bridges with the road surface quality on one being described as poor and on the other as good. The test vehicles were run across the bridges at 10 and 40 mph. Besides providing DAFs for the different vehicles on the different road surfaces, the tests indicated that the DAFs derived from the theoretical analyses were realistic. They also confirmed that the dynamic response of a bridge to a moving load could be determined realistically by a static analysis using the measured but varying wheel loads. The measurements of the dynamic wheel forces from the tests carried out on the three different road surfaces showed that significant reductions in the bridge response as compared to the normal were only to be expected from a 'good' road surface. Thus only two categories of road surface were considered in deriving the load reduction factors to be included in the revision of the assessment loading.

8.5 Probabilistic load model

The next stage was the development of a probabilistic load model for use in the reliability analysis which was to be used to calibrate the revised loading proposals. The model was composed of the following three elements:

(i) a model based on Load Model 1 in *Eurocode 1 Part 3* [4] to give a set of static load effects;
(ii) a statistically defined factor to enable the required probability distribution of the load effects to be derived from the basic static load model;
(iii) a statistically defined factor to enable the static load effects to be enhanced to take account of the dynamic effects.

As already described, the parameters of the statistical factors were derived from the M6 traffic data and the dynamic tests for different situations which meant that the resulting statistical load models could represent different traffic flows and road surfaces for the various spans being considered. Separate statistical load models were derived from simply-supported bridge studies to cover the following cases:

(i) extreme lane 1 effects due to the passage of a single very heavy vehicle;
(ii) joint extreme multi-lane effects due to the coincident passage of heavy vehicles;
(iii) heavily compressed stationary traffic, closely spaced laterally.

8.6 Reliability analysis

Since the existing assessment load model consisted of the product of a formula loading and partial factors it was felt necessary to investigate whether the partial factors should be varied for the less than onerous cases in order to have a consistent notional reliability for the different road and traffic situations. Accordingly a limited number of reliability analyses were carried out by Imperial College [5]. Simply supported reinforced concrete slab bridges and reinforced concrete beam and slab bridges with spans up to 20 m were investigated. The fictitious bridges were designed using the loads specified in BD 21/93 [6] and the relevant strength clauses from the appropriate part of BS 5400 [7]. Optimum sets of partial factors were obtained for the six scenarios of traffic and road conditions with the live-load factor then being adjusted to obtain uniform reliability.

In the case of slab bridges there was little variation in the live-load factor for the different traffic flows, but the partial factors required for the poor road surfaces were on average 7.5 per cent greater than those for the good surfaces. This implied that the assessment load for bridges with good road surfaces should be about 7 per cent below that for a poor surface. In the case of the beam and slab bridges the live load partial factors for the good surface cases were generally lower than for the poor surfaces. However, there was approximately a 6 per cent increase in the partial factors required for the low traffic flow cases. These findings were used to modify the load reduction factors *K* obtained from the characteristic load effect comparisons.

8.7 Lateral bunching

In the description of the derivation of the short-span loading in Chapter 5 it will be noted that there was assumed to be no correlation between the allowance for impact and the allowance for the lateral bunching of vehicles. The rule was that the impact allowance was only to be applied to a single vehicle when it provided the worst load case. The lateral bunching was allowed for by a span dependent factor which increased the loading on the shorter spans and which depended also on the width of the notional lane. During the development of the assessment code it was suggested that there should be a single factor to cover these two cases, since it was felt to be unlikely that two vehicles travelling closely side by side would be travelling fast enough to generate any significant dynamic effects. However, the working party decided that even though the possibility of vehicles travelling closely side by side at speed was remote it was nevertheless possible, and that the assumption was therefore very much in accord with the concept of determining the most extreme load that could reasonably be expected to occur in the lifetime of the bridge.

The problem of lateral bunching was reconsidered during the review of short-span loading described in this chapter. It was pointed out that although the maximum width of the bodies of heavy goods vehicles was 2.5 m, they had large wing mirrors which increased their effective overall width to nearly 3.0 m. As a consequence they would be unlikely to drive very closely side by side so as to avoid damaging their mirrors. It was considered therefore that tight lateral bunching in normal traffic would be rare and that the load model should therefore cater for either moving vehicles with dynamic effects but no lateral bunching, or stationary vehicles with lateral bunching but no dynamic effects.

These two alternatives were investigated by comparing characteristic bending moments. It was found that, even assuming the most onerous conditions for the stationary bunched traffic, the load effects from the flowing traffic without lateral bunching exceeded those from jammed traffic with bunching. It was concluded that no lateral bunching allowance was needed and that any loading derived from the assessment code assuming a 2.5 m lane width would be satisfactory for all notional lane widths.

8.8 Load reduction factors

Using the probabilistic load models, characteristic load effects were derived for a range of spans and numbers of lanes and for the six traffic scenarios. Preliminary values of a load adjustment factor *K* were then obtained, *K* being defined as the ratio between the characteristic effect obtained using the relevant probabilistic load model and the effect obtained using the HA loading for a 2.74 m lane width specified in BD 21/93 [6]. These *K* factors were then multiplied by a constant normalising factor to ensure compliance with the Highways Agency's decision

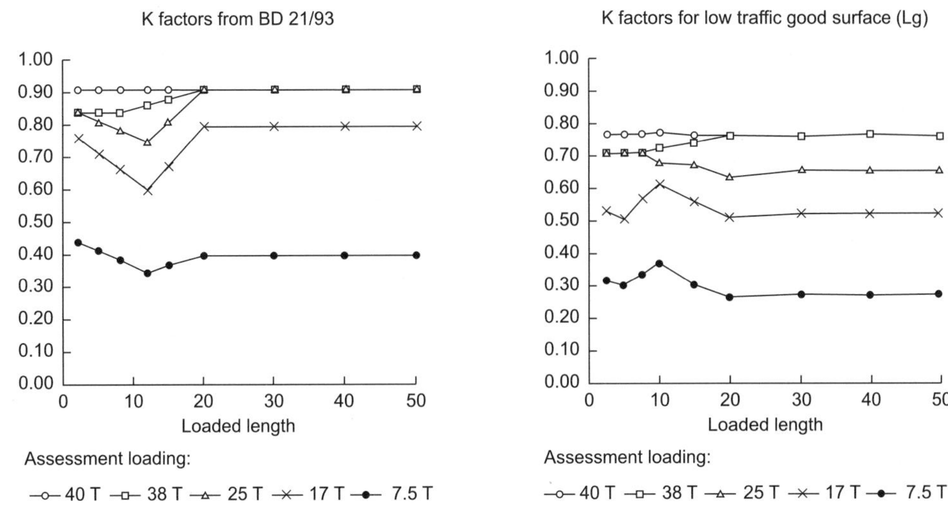

Figure 8.2. Comparison of load reduction factors for low traffic flow on a good surface [2]

that there was to be no change in the assessment loading for the most onerous conditions. Thus the loading for the scenario representing heavy traffic flows on a bridge with a relatively poor road surface were adjusted to be similar to the '40 tonne' assessment loading given in BD 21/93 [6] by means of the normalising constant.

The load effects derived from the M6 traffic data were taken as being representative of the '40 tonne' assessment loading and the reduction factors for other levels of assessment loading (e.g. 25 tonnes, 17 tonnes) were calculated by reference to this case. Values for the 38-tonne level were obtained deterministically by factoring the 40-tonne values by the ratios between the 38- and 40-tonne values given in BD 21/93 [6]. Reduction factors for the 25-tonne level were obtained by repeating the static load effect simulations with all vehicles with more than 3 axles being filtered out of the M6 data. Similarly the reduction factors for the 17-tonne level were obtained by filtering out all vehicles with more than 2 axles from the original data. The reduction factors for the 7.5-tonne level were obtained deterministically by using the BD 21/93 [6] ratios as before. Figure 8.2 shows the load reduction factors obtained from the probabilistic study for a low traffic, good road surface situation compared with the assessment live load factors given in BD 21/93 [6]. In general it was found that the reductions in assessment load which were available for the good road surface compared to the poor surface were negligible for the lower assessment levels at the longer spans because of the stationary traffic situation dominating the load effects.

The values of the accidental axle and wheel loads for the different levels of assessment loading and for the six traffic scenarios were obtained by factoring the maximum values, 220 kN and 100 kN for a single axle and a single wheel respectively, in the same proportions as the relevant load reduction factors for the shortest span listed. The live-load reduction factors for the UDL, KEL and accidental loads were later incorporated into a revision of the assessment code which was issued in 1997 as BD 21/97 [8].

References

1. Department of Transport (1984) Departmental standard BD 21/84. *The assessment of highway bridges and structures*. London: DTp.
2. Flint and Neill Partnership (1996) *Probabilistic assessment of short span bridges*. Transport Research Laboratory Contractors report CR16. Unpublished project report for Highways Agency.
3. Ricketts, N. J. and Page, J. (1997) Transport Research Laboratory report 251, *Traffic data for highway bridge loading*. Crowthorne: TRL.
4. European Committee for Standardisation (1994) ENV 1991-3, *Eurocode 1: Basis of design and actions on structures—Part 3: Traffic loads on bridges*. Brussels: CEN.
5. Chryssanthopoulos, M. K. and Micic, T. V. (1996) *International symposium on the safety of bridges—Reliability evaluation of short span bridges*. Thomas Telford Conferences, London.
6. Department of Transport (1993) *DMRB*, Vol. 3, Section 4, Part 3: BD 21/93, *The assessment of highway bridges and structures*. London: DTp.
7. British Standards Institution (1980) BS 5400: Parts 1—10, *Steel, concrete and composite bridges*. London: BSI.
8. Highways Agency (1997) *DMRB*, Vol. 3, Section 4, Part 3, BD 21/97, *The assessment of highway bridges and structures*. London: HMSO.

Chapter 9
Loading for assessment

9.1 Introduction

The previous chapters have dealt mainly with the live traffic loading to be applied for the design of new structures. However, for the bridge owner it is just as important to be able to assess the safe carrying capacity of an existing structure in terms of the vehicles which should, or should not, be permitted to use the bridge in question. This is particularly relevant for structures which have been built to earlier, and possibly less demanding, standards, and for structures which are showing signs of significant deterioration. Although design codes could be used for this purpose they are usually not entirely suitable for use in assessment. They do not always cover the older methods of construction and because they only contain one level of loading it is difficult to ascertain the actual level of loading which could be carried by a structure which has been found to be sub-standard. Thus there has been a need, which has grown as the stock of bridges in the UK itself has grown and aged, for dedicated codes for bridge assessment which have their own specific live-loading criteria.

In the period after the Second World War there was increasing concern about the condition of the UK stock of highway bridges, and it was realised that there was a need for a common and consistent approach to their assessment. Before and during the war the Building Research Station had undertaken an extensive investigation [1] into the behaviour of cast iron and masonry arch bridges under load in order to find a basis for assessing their load carrying capacity. The results were used as the basis of the assessment and classification of bridges on major routes for their ability to carry certain military vehicles. The results of the work also assisted in the development of the MEXE method [2] for assessing masonry arch bridges which was later developed for civilian use. Further studies of bridge behaviour were done by the Building Research Station after the war [3], but it was not until the 1960s that the first comprehensive bridge assessment code, *Technical Memorandum (bridges)* No. BE 4

[4], was produced by the MoT using much of the information from the studies which have been described.

Although it is a feature of this, and later, codes that they contained guidance on methods of analysis, including simple methods of load distribution for certain forms of construction, this chapter will focus on the main loading criteria and their application. It will also deal with the way that any restrictions on traffic loading were to be determined for structures which were found to be incapable of carrying the full range of vehicles permitted under the *C&U regulations*.

9.2 Technical Memorandum (bridges), BE 4 [4]

9.2.1 Background

Technical Memorandum BE 4 was written by the Ministry of Transport and the British Railways Board in consultation with the County Surveyors Society. When initially published in 1967 it consisted of two parts, namely Part I, *Basic principles*, and Part II, *Quick assessment of bridges containing longitudinal or transverse members*. A third part, Part III, *Assessment of masonry and brick arches*, was added later. The code was intended for the assessment for C&U vehicles of bridges which were not capable of carrying the HA loading as specified in BS 153: Part 3A [5]. The standards adopted in the code, including the traffic loading, were lower than those used in the design of new bridges, because it was assumed that the replacement of any sub-standard structures would not be unduly delayed.

9.2.2 Vehicle trains

Unlike the equivalent formula type loading in the design code BS 153: Part 3A [5], the loading in BE 4 was in the form of two trains of idealised vehicles, namely a train of five 2-axle vehicles and a train of three 4-axle vehicles. The axle weights and

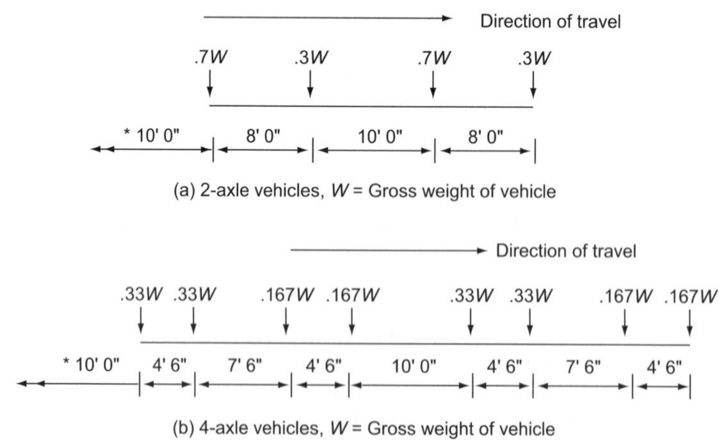

Figure 9.1. Assessment loading trains in BE 4

spacings for both trains are shown in Fig. 9.1. It will be noted that the gross weight of each vehicle is expressed as an unknown, W, since the maximum value was to be determined from the analysis of the structure and the strength of the various structural members. Graphs were included in the code which gave the values of the bending moments and shears in longitudinal members generated by trains of vehicles of different gross weights for a range of spans up to 100 ft. There were also graphs giving the moments on cross-girders from a single axle and a pair of axles side by side for different axle weights and for spans up to 40 ft.

Although the loading trains and the accompanying load effect graphs were derived for vehicles permitted under the 1963 *C&U regulations* [6] with a maximum weight of 24 tons on 4 axles, they in fact covered the heavier vehicles allowed under the 1964 regulations which increased the maximum permitted weights to 28 tons for 4-axle rigid vehicles and 32 tons for 4-axle articulated vehicles. This was allowable because the increases in gross weights had been accompanied by restrictions on the minimum axle spacings of the vehicles concerned, which ensured that any increases in loading effects were only marginal. It was pointed out that the effects of a train of 14-ton 2-axle vehicles were as great as those of any other permitted vehicle with a gross weight of up to 20 tons. Similarly for the 4-axle vehicle train the effects of the 24-ton vehicle were said to be greater than the effects from all the other vehicles complying with the *C&U regulations*, although some adjustments were made to the 24-ton curves in the graphs to allow for the effects of a single 11-ton axle and a twin 9-ton axle.

9.2.3 Application of vehicle trains

For longitudinal bending either of the two vehicle trains could be applied as appropriate, whereas for longitudinal shear only the 4-axle vehicle train was to be applied. For transverse members the choice of loading was between the 4-axle vehicle train and a single 11-ton axle. Only one train of vehicles was to be applied for carriageway widths less than 18 ft and not more than two trains for widths greater than 18 ft. There was an exception for crowded traffic situations, such as at traffic lights at the end of a bridge, where a greater number of trains could be considered. Where two trains of vehicles were being considered the transverse positions of the vehicle tracks were to be taken as shown in Figure 9.2.

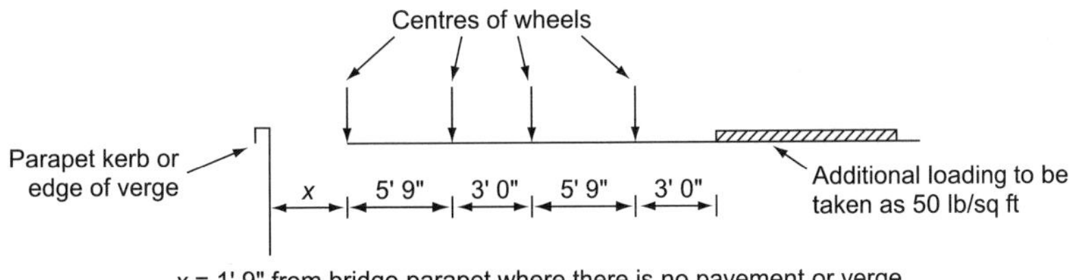

x = 1' 9" from bridge parapet where there is no pavement or verge
x = 1' 9" from kerb where there is a pavement
x = 9" from edge of verge where there is no kerb

Note: Most commercial vehicles have track centres of approx. 5' 9" and an overall track of 7' 9".

Figure 9.2. Positioning of vehicle trains [4]

Where the length of the bridge in question exceeded the length occupied by the vehicle train, a UDL of 50 lb/ft^2 was to be applied to the remainder of the unoccupied length of the deck. A UDL of 50 lb/ft^2 was also to be applied to the excess width of wide bridges capable of carrying more than two trains when considering the loading on abutments and cross-girders. Impact was to be allowed for by applying a 25 per cent increase to the load on one axle in one train at the position having the greatest effect on the member being considered. The contact area for a wheel was to be obtained by allowing 33 in^2 for each ton carried by the wheel.

9.2.4 Public service vehicles

Special allowances were made for public service vehicles (i.e. buses and coaches) because for a given gross weight they tended to have longer wheelbases than the minimum allowed under the *C&U regulations*. Thus a bus could be heavier than a vehicle in a loading train but have no greater bending moment effect. A table was included in BE 4 which gave the equivalent weights of the 2-axle assessment vehicle for a number of different makes of public service vehicle (PSV) for a range of spans. For example a 13.2-ton AEC bus was equivalent to a 10-ton assessment vehicle on a 40 ft span.

9.2.5 Load distribution

Technical Memorandum BE 4 gave simple methods for determining the distribution of bending moments and shears in certain older forms of construction such as longitudinal beams with jack arches. These methods used graphs of proportion factors for the different members derived from the theoretical work of Hendry and Jaeger [7] and the test results from model and full-size bridges published by Thomas and Short of the Building Research Station [3].

9.2.6 Operation bridgeguard

Technical Memorandum BE 4 was used as the basis for assessing, and strengthening where necessary, all the older bridges on trunk roads under a programme called 'Operation bridgeguard'. Subsequently the programme was extended to cover the bridges carrying roads over railways and waterways owned by British Rail and British Waterways respectively. Although the use of the code was recommended to other bridge owners, such as local authorities, it appears that only a small proportion of their bridges were assessed at the time. The major part of the work under Bridgeguard was carried out in the 1970s although some work was still being done in the 1980s.

9.3 Technical Memorandum (bridges) BE 3/73 [8]

Technical Memorandum BE 4 was superseded by the issue by the Department of the Environment of BE 3/73 in 1973. The loading for bridges with slab-type decks was still specified in the form of the two trains of 2-axle and 4-axle vehicles. The main change was the inclusion of references to changes in the 1972 *C&U regulations* which increased the maximum weight of 4-axle rigid vehicles to 30 tons. But here again the minimum axle spacings of the vehicles were controlled so that the effects of these vehicles were no greater than the 24-ton vehicles. Thus the various graphs of loading effects derived from trains of vehicles up to 24 tons in weight were still applicable to all the vehicles permitted under the *C&U regulations*. Similarly, although two closely spaced axles of up to 10 tons were permitted, their spacing was such that their effect was equivalent to the effect of two 9-ton axles at a slightly closer spacing of 4.5 ft. As in BE 4 the strength calculations for the different elements and materials were based on permissible stresses and there were no factors to be applied to the loading.

9.4 Departmental standard BD 21/84 [9]

9.4.1 Background

The background to the development of a new assessment code, BD 21/84, which was based on limit state principles, has been partly covered in Chapter 5. In particular the derivation of the Type HA loading curve covering normal vehicles has been described in detail. This chapter will therefore focus on the way that the loading was to be applied, the derivation and use of the reduced loadings, and other secondary loading criteria. One of the aims of the new code was to ensure that the safety margins implicit in the assessment process were of the same order as those adopted for the design process, because it was realised that there could be no guarantee that any sub-standard structures would soon be replaced. This led to the development of a traffic model which was felt to be more representative of the traffic of the day than the idealised loading trains in BE 4 and BE 3/73, and which made allowance for overloading, lateral bunching and higher impact forces.

9.4.2 Reduced loadings

One of the differences between the traffic loading models used for design, as for instance given in BD 37/88 [10], and those used for assessment is that in addition to the full standard loading, covering the full range of vehicles permitted under the *C&U regulations*, the assessment code also included so-called 'reduced loadings'. The values of these reduced loadings were determined in a similar way to the standard loading as described in Chapter 5, but used a C&U envelope of loading effects which only included those vehicles whose gross weights were equal to or less than the maximum weight ascribed to the level of loading being derived.

However, no 10 per cent contingency margin was included in the calculations for the reduced loadings although it was still included in the full loading. The purpose of the reduced loadings was to help determine a suitable vehicle weight limit for those structures which were found to be unable to carry the full standard loading. There were three main levels of loading representing particular groups of general commercial vehicles, three subsidiary levels representing public service vehicles and fire engines and one level representing cars and light vans. Details of the different levels of reduced loading are as follows:

(i) Level One, representing the loading from all types of 2- and 3-axle C&U vehicles with a maximum weight of 24.5 tonnes.

(ii) Level Two, representing the loading from all types of 2-axle C&U vehicles with a maximum weight of 16.5 tonnes.

(iii) Level Three, representing the loading from 2-axle light goods vehicles with a maximum weight of 7.5 tonnes.
(iv) PSVs and fire engines (FEs), representing the loading from specified makes of buses, coaches and FEs, details of which were included in an appendix. The loading allowed for up to three vehicles in convoy with the remaining space being filled with car loading. No overload factor was included for fire engines.
(v) Cars, representing the loading from cars and light vans with a maximum weight of 3 tonnes.

The reduced loadings for a particular span were obtained by multiplying the standard UDL and the 120 kN KEL by the appropriate reduction factor from Fig. 9.3.

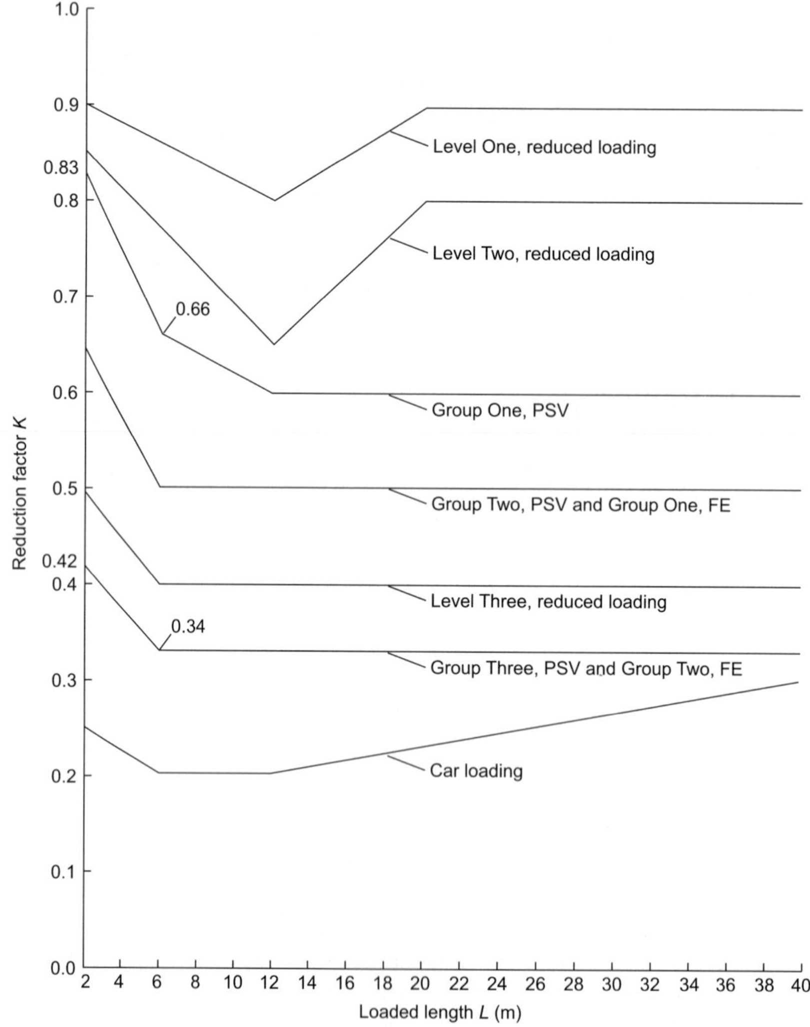

Figure 9.3. Reduction factors for UDL and KEL of reduced loadings [9]

9.4.3 Single axle and single wheel loads

In addition to the UDL and KEL, a single axle and a single wheel load could also be applied. These were to be applied instead of the UDL and KEL for spans below 2 m and also for the assessment of non-trafficked areas such as central reserves and footways. A table gave the values of the nominal axle and wheel loads for the different loading levels: for the standard loading these were 200 kN and 100 kN respectively. The wheel loads were to be distributed over a circular or square area to give a pressure of 1.1 N/mm^2.

9.4.4 Application of loading

In general BD 21/84 [9] only required structures to be assessed for Type HA live loading as represented by the UDL and KEL. There was no requirement to apply Type HB loading although a requirement to check the number of HB units that a structure could carry was later introduced into one of the documents [11] giving instructions for carrying out the 15-year bridge rehabilitation programme. Special loadings were given for certain types of construction such as trough decks, short-span masonry arches and decks with transverse members. The last were incorporated in a complementary Advice note BA 16/84 [12] and were necessary because it was considered that loads on transverse members were not modelled satisfactorily by the equivalent UDL and KEL. Departmental standard BD 21/84 [9] only required the assessments to be carried out for the ULS for which a partial live load factor of 1.5 was to be used. There were no specific requirements for load combinations apart from dead plus live load, although if it were thought to be necessary to consider other loads reference was to be made to BS 5400: Part 2 [13].

As in the case of design (see Chapter 7), the carriageway was to be divided into a number of notional lanes with a minimum width of 2.5 m. For carriageways less than 5.0 m wide only one notional lane of 2.5 m was to be taken with the remaining width being loaded with a UDL of 5 kN/m^2. Because the lateral bunching included in the derivation of the UDL assumed a lane width of 3.65 m, the UDL was to be reduced by a lane width factor for notional lanes which were less than 3.65 m wide. In BD 21/84 [9] the lane width factor was given in graphical form as shown in Fig. 9.4 rather than as a formula. The full appropriate UDL and KEL were to be applied to up to two notional lanes with 0.6 times the UDL and KEL being applied to all other lanes. However, there was a proviso in BD 21/84 [9] that because the shear case always determined the value of the equivalent UDL by a margin of about 10 per cent, the longitudinal bending moment only could in certain cases be reduced by 10 per cent, but this was deleted from later versions.

9.4.5 Centrifugal effect

Departmental standard BD 21/84 [9] gave requirements for calculating the vertical effects arising from the centrifugal forces on a horizontally curved carriageway together with a full derivation of the various equations. These made use of the work by Burt at the Transport Research Laboratory (TRL) [14]. The calculations involved deriving equivalent static longitudinal line and point loads from the relevant UDL which represented the wheels of lorries. The loads on the outside wheels were enhanced by a centrifugal effect factor whilst those on the inside of

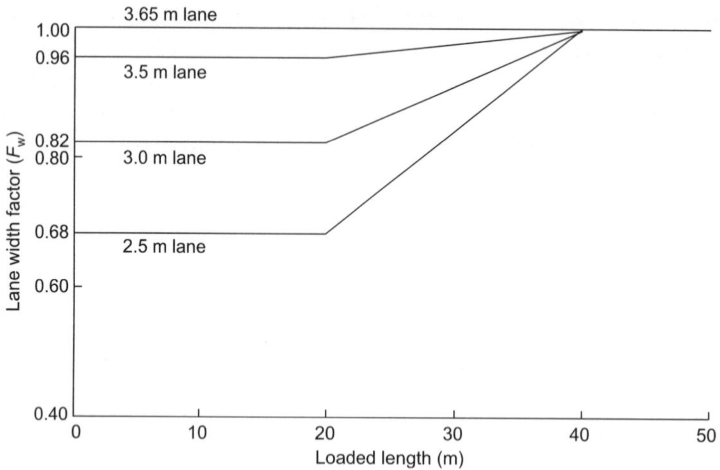

Figure 9.4. Lane width factor [9]

the curve were reduced. Centrifugal effects could be ignored when the horizontal radius of curvature exceeded 600 m, the span of the element under consideration was greater than 15 m or the bridge had a reinforced or prestressed concrete slab deck. It will be noticed that there was no mention of the need to take account of the lateral effect of centrifugal forces on the bridge as a whole.

9.5 Advice note BA 16/84 [12]

The issue of BD 21/84 [9] was accompanied by the issue of a complementary Advice note BA 16/84. This was mainly concerned with the simple methods of load distribution for certain types of construction using the load distribution factors which had been given in BE 3/73 [8]. However, it also contained graphs and tables of equivalent axle loads for the direct determination of load effects in particular forms of deck construction and reduction factors to calculate reduced loadings in transverse members.

In addition BA 16/84 included the empirical method for the load assessment of masonry arch bridges based on the MEXE method [2]. Of interest, from the loading point of view, were the graphs which allowed the double-axle loads inherent in the method to be converted to equivalent single and triple axles. The graphs also dealt with so called 'axle lift-off' where one axle of a double or triple axle bogie could lift off from the road surface and transfer extra load to the other axles and affect the loading on the arch.

9.6 Amendment No. 1 to BD 21/84 [15]

9.6.1 Background

In 1989 the Department of Transport issued Amendment No. 1 to Departmental standard BD 21/84. Apart from some corrections and clarifications the main changes were concerned with the loading requirements. These were amended to cover the effects of a set of vehicles being proposed by the European Commission which included a 40-tonne vehicle and 11.5-tonne drive axles. These vehicles were additional to the vehicles in the *C&U regulations* which were already covered including the 38-tonne vehicle and the 10.5-tonne drive axle. The assessment loading was thus brought into line with the design loading for spans up to 50 m, which was included in the composite version of BS 5400: Part 2 described in Chapter 7, and it also included the 10 per cent contingency margin. What had been known as the standard loading became the Type HA loading with the uniformly distributed loading curve having the same formula as the design curve. The HA UDL values were to be multiplied by appropriate reduction factors to give the different levels of assessment live loading. In reviewing the assessment loading it was felt that the 10 per cent contingency allowance should no longer be applied to the reduced levels of loading as structures which were found to be sub-standard would be earmarked for strengthening or replacement under the Department's 15-year bridge rehabilitation programme. Thus any assessment loading determined from the HA loading curve no longer contained a 10 per cent margin.

9.6.2 Assessment loading

As a result of the changes outlined above a new set of reduction factors were produced for the different levels of assessment loading. These are shown in Fig, 9.5 where it will be noted that the special loading criteria for PSVs were no longer included. This was because there had been such a dramatic increase in the types of buses and coaches being built and operated in the UK, that it was no longer practical to maintain an up-to-date list of such vehicles or to examine their individual loading effects. Because of the inclusion of the EC vehicles and some slight changes in the *C&U regulations* the descriptions of the assessment live loadings (ALLs) were amended to the following:

- (i) 40 tonnes representing the full range of EC and C&U vehicles up to 40 tonne gross vehicle weight (gvw) and 11.5-tonne drive axle.
- (ii) 38 tonnes representing the full range of vehicles complying with the current *C&U regulations*.
- (iii) 25 tonnes representing the loading from all types of 2- and 3-axle vehicles including PSVs up to 25 tonnes gvw.
- (iv) 17 tonnes representing the loading from all types of 2-axle vehicles including PSVs up to 17 tonnes gvw.
- (v) 7.5 tonnes representing the loading from all types of 2-axle light goods vehicles and PSVs up to 7.5 tonnes gvw.
- (vi) FE representing the loading from two groups of makes of fire engine listed in the code.
- (vii) 3 tonnes representing the loading from all types of car and van.

9.6.3 Wheel loading

Slight changes were made to the table giving the values of single axle and wheel loads to correspond to the definitions of the assessment live loading levels, and the nominal single axle and wheel loads of 200 kN and 100 kN were now associated

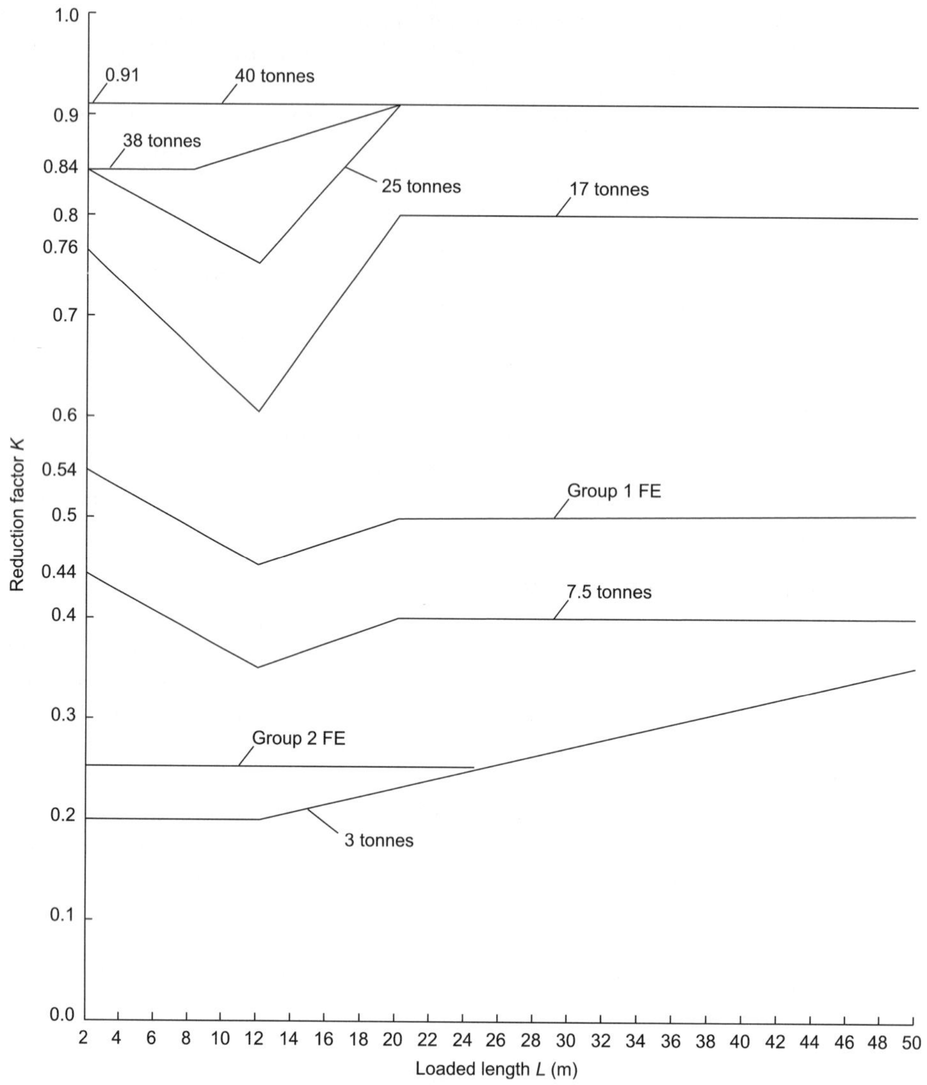

Figure 9.5. Reduction factors for UDL and KEL of assessment live loadings [15]

with the 40-tonne ALL rather than the standard loading. However, a more significant change was the introduction of accidental loading for members supporting a non-trafficked area of the deck which was not protected by an effective barrier. The values of the nominal accidental wheel loads for the different loading levels are given in Table 9.1; the arrangement of the four wheels on two axles was the same as in the composite vérsion of BS 5400: Part 2 [13].

9.6.4 Special structures

Attention was again drawn to the shortcomings of the Type HA loading in modelling the effects of vehicles on certain types of structure now including

Table 9.1. Nominal accidental wheel loads [15]

Assessment live loading	W_1 (kN)	W_2 (kN)	a (m)
40 tonne	100	60	1.5
38 tonne	90	60	1.5
25 tonne	90	40	1.5
17 tonne	90	10	1.5
7.5 tonne	50	10	1.5
3 tonne	25	–	–
FE Group One	60	10	1.5
FE Group Two	30	20	1.5

buried concrete boxes. For this type of construction (e.g. cross-girders and transverse spanning slabs) the method of loading using individual vehicles was given in an appendix. This included a list of critical vehicles to be used in assessment with details of their axle weights and spacings.

9.6.5 Lane factors

Another change to BD 21/84 to bring it into line with the design code was the adoption of the same table of lane factors as in the composite version of BS 5400: Part 2 [13]. These encompassed both the lane width factor and the values of the HA UDL and KEL to be applied to the various lanes in a multiple-lane carriageway.

9.7 *Design manual for roads and bridges* BD 21/93 [16]

The amendments listed above were later incorporated into a complete revision of BD 21 which was issued as part of the *Design manual for roads and bridges* in 1993. At the same time an updated version of the Advice note was issued as BA 16/93 [17]. This contained an updated version of the table relating the maximum allowable gross weights of vehicles on masonry arches in terms of their assessed allowable axle loads. The figures giving the factors for converting the calculated allowable double-axle loads to other axle configurations were amended and simplified in the light of some recent research regarding the transfer of axle loads in tri-axle bogies at 'lift-off'.

9.8 *Design manual for roads and bridges* BD 21/97 [18]

This version of BD 21 was issued in 1997 and incorporated the results of the work described in Chapter 8 to derive assessment loading which was site specific and took account of traffic flows and the condition of the road surface. The basic HA loading curve remained the same as before covering the C&U vehicles and the proposed EC vehicles. The latter now included vehicles up to 44 tonnes gross weight but these were long vehicles used in combined road/rail operations and certain drawbar trailer combinations. The single set of reduction factors to obtain the reduced levels of assessment loading were now replaced by six sets of factors to cover six traffic scenarios representing heavy, medium and low traffic flows and poor and good

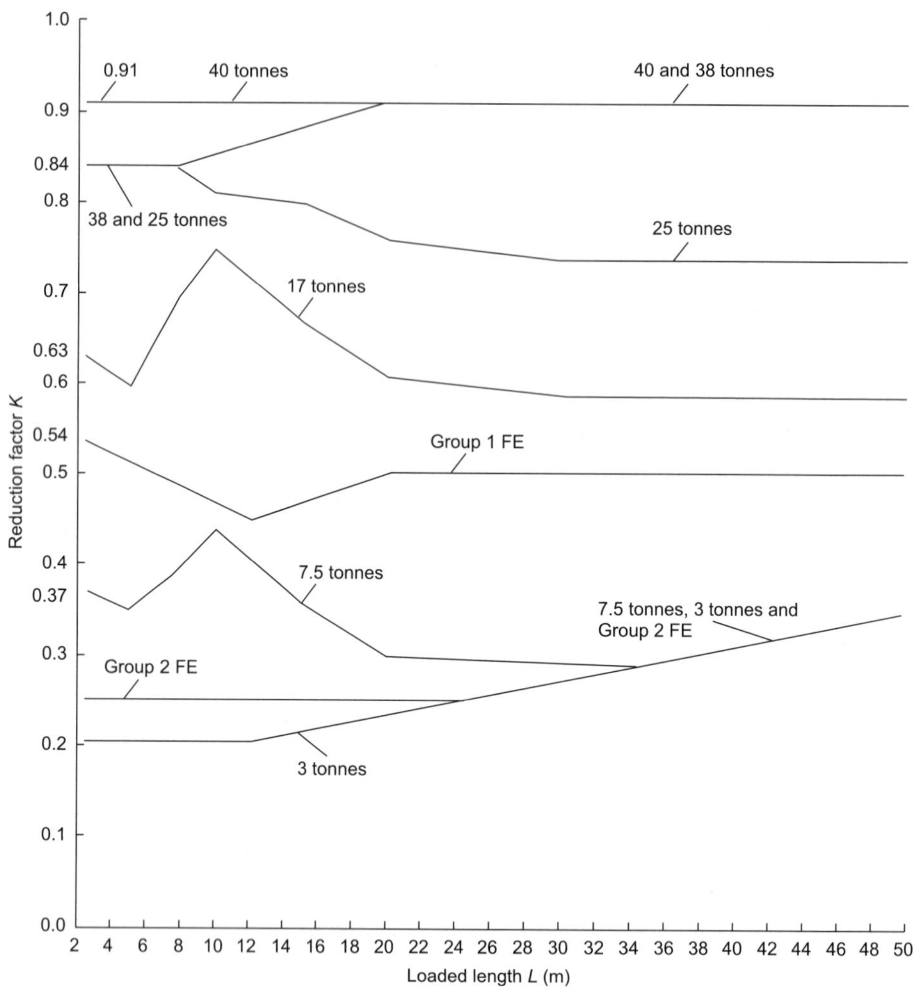

Figure 9.6. Reduction factors for heavy traffic poor surface [18]

road surfaces. A typical example of the reduction factors for heavy traffic on a poor road surface is shown in Fig. 9.6. The definitions of the different traffic flow categories and road surface conditions were given in the code. The six traffic scenarios also applied to the values of the nominal single-axle and single-wheel loads for the different levels of assessment loading. Thus for the 25-tonne level the nominal single-axle load for a medium traffic flow on a good road surface is 160 kN compared to the value of 180 kN in BD 21/93 [16].

9.8.1 Adjustment factors

In the derivation of the short-span live loading, described in Chapter 5, a lateral bunching factor was included to take account of the possibility that, in slow moving traffic, the lateral gap between two lines of vehicles could be reduced resulting in more lines of traffic on a bridge than indicated by the number of marked or notional lanes. This lateral bunching effect was assumed to occur with high-impact effects, but later work, described in Chapter 8, using a probabilistic approach to the derivation of short-span loading, showed that the two effects should not be considered to be acting together and that the high-speed high-impact effect with no lateral bunching produced the most onerous loading. Accordingly, it was decided that the lateral bunching effect should be eliminated from the HA UDL and KEL. Rather than re-derive the equivalent uniformly distributed loading curve the adjustment was to be done by dividing the relevant value of the UDL and the 120 kN KEL by an adjustment factor whose value was determined in a similar manner to the determination of the lateral bunching factor, as described in Chapter 5, but assuming a notional lane width of 2.5 m in every case.

9.8.2 Special structures

Requirements for the loading of certain types of structure using critical vehicles were again included in an appendix together with the list of vehicles. However, the list was enlarged by the inclusion of the critical EC vehicles, up to 40 tonnes, and a list of the C&U vehicles to be used when assessing for reduced levels of loading.

9.8.3 Lane factors

Following the work described in Chapter 8, and the decision to eliminate lateral bunching it was necessary to provide explicit lane factors for the live loads to be applied to the various lanes of a multiple-lane carriageway. The work in Chapter 8 influenced the choice of values for the lane factors which were slightly different from those in earlier versions of the code, namely:

Lane 1—1.0
Lane 2—1.0
Lane 3—0.5 and
Lane 4 and subsequent lanes—0.4.

It was specified that irrespective of the notional lane width the UDL and KEL should occupy a width of 2.5 m in the most onerous position in that lane. This was to allow for the possibility of the loading on a longitudinal beam of a beam and slab bridge from the offside wheels of one vehicle and the nearside wheels of a vehicle in an adjacent lane producing a more onerous effect than if the equivalent UDL + KEL had been spread over the full width of a wider notional lane.

9.9 *Design manual for roads and bridges* BA 16/97 [19]

A new version of the complementary Advice note was issued in 1997 with no significant changes regarding any live loading criteria. An amendment to the Advice note was issued in 2001 to incorporate changes in the allowable vehicle and axle weights promulgated in the *Road vehicles (authorised weight) regulations* 1998 [20]. These took on board the maximum weights for certain vehicles and axles which had been proposed by the European Commission. The changes affected the assessment live loading levels and the allowable gross vehicle weights for arches derived from the allowable axle loads.

9.10 *Design manual for roads and bridges* BD 21/01 [21]

A revised version of BD 21 was issued in 2001 and incorporated the changes in vehicle and axle weights noted above which had been promulgated in the *Road vehicles (authorised weight) regulations* 1998 [20]. Whereas previously the proposed EC vehicles had been considered separately from those permitted under the C&U regulations they were now treated as one set. The maximum allowable vehicle weight had increased to 44 tonnes on 5 axles, for vehicles used in intermodal (i.e. rail/road) journeys, and to 44 tonnes on 6 axles, but with 10.5-tonne drive axles.

One of the main changes was to the definitions of the assessment live load levels which were now defined as follows:

- (i) 40 tonnes covering the full range of vehicles up to 40/44 tonnes gvw.
- (ii) 26 tonnes representing the loading from all types of 2- and 3-axle vehicles up to 26 tonnes gvw.
- (iii) 18 tonnes representing the loading from all types of 2-axle vehicles up to 18 tonnes gvw.
- (iv) 7.5 tonnes representing the loading from all types of 2-axle light goods vehicles and PSVs up to 7.5 tonnes gvw.
- (v) FE loading representing the loading from the two groups of FEs listed in the code.
- (vi) 3 tonnes representing the loading from cars and vans up to 3 tonnes gvw.

The graphs of the reduction factors for the six scenarios of traffic flow and road surface condition were rationalised to match the revised assessment load level definitions with some amendments to the plotted values. Similar adjustments were made to the tables giving details of the critical vehicles to be used in the analysis of certain structures or when looking at transverse effects. Adjustments were also made to the descriptions of the assessment load levels for the single axle, single wheel and accidental wheel loads, with some adjustment of the nominal values for the six different traffic scenarios.

References

1. Davey, N. (1953) National Building Studies: Research paper No. 16. *Tests on road bridges*. London: HMSO.
2. Military Engineering Experimental Establishment (1963) *Classification (of civil bridges) by the reconnaissance and correlation methods*. Christchurch: MEXE.
3. Thomas, F. G. and Short, A. (1952) 'A laboratory investigation of some bridge deck systems', *Proceedings of Institution of Civil Engineers*, March.
4. Ministry of Transport (1967) Technical Memorandum (bridges) No. BE 4, *The assessment of highway bridges for Construction and Use vehicles*. London: MoT.
5. British Standards Institution (1954) BS 153: Part 3A: 1954, *Girder bridges Part 3: Loads and stresses—Section A: Loads*. London: BSI.

6. Statutory Instrument 1963 (1963) *The motor vehicles (construction and use) regulations 1963* as amended. London: HMSO.
7. Hendry, A. W. and Jaeger, L. G. (1958) *The analysis of grid frameworks and related structures*. London: Chatto and Windus (reprinted 1969).
8. Department of the Environment (1973) Technical Memorandum (bridges) No. BE 3/73, *The assessment of highway bridges for Construction and Use vehicles*. London: DoE.
9. Department of Transport (1984) Departmental standard BD 21/84, *The assessment of highway bridges and structures*. London: DTp.
10. Department of Transport (1989) Departmental standard BD 37/88, *Loads for highway bridges*. London: DTp.
11. Department of Transport (1988) Departmental standard BD 34/88, *Technical requirements for the assessment and strengthening programme for highway structures. Stage 1—older short-span bridges and retaining structures*. London: DTp.
12. Department of Transport (1984) Advice note BA 16/84, *The assessment of highway bridges and structures*. London: DTp.
13. British Standards Institution (1978) BS 5400: Part 2: 1978, *Steel, concrete and composite bridges—Part 2. Specification for loads*. London: BSI.
14. Burt, M. E. (1969) Road Research Laboratory Technical Note No. TN 401, *Forces on bridges due to braking vehicles*. Unpublished report.
15. Department of Transport (1989) Amendment No. 1 to Departmental standard BD 21/84, *The assessment of highway bridges and structures*. London: DTp.
16. Department of Transport (1993) *DMRB*, Vol. 3, Section 4, Part 3, BD 21/93, *The assessment of highway bridges and structures*. London: DTp.
17. Department of Transport (1993) *DMRB*, Vol. 3, Section 4, Part 4, BA 16/93, *The assessment of highway bridges and structures*. London: DTp.
18. Highways Agency (1997) *DMRB*, Vol. 3, Section 4, Part 3, BD 21/97, *The assessment of highway bridges and structures*. London: HMSO.
19. Highways Agency (1997) *DMRB*, Vol. 3, Section 4, Part 4, BA 16/97, *The assessment of highway bridges and structures*. London: HMSO.
20. Statutory Instrument 1998 (1998) No. 3111. *The road vehicles (authorised weight) regulations 1998*. London: HMSO.
21. Highways Agency (2001) *DMRB*, Vol. 3, Section 4, Part 3, BD 21/01,. *The assessment of highway bridges and structures*. Norwich: HMSO.

Chapter 10
Abnormal loading

10.1 Introduction

Much of the work described in the earlier chapters has been concerned with the development of loading models for 'normal' traffic. A great deal of work has been done to ensure that this Type HA loading has been kept up to date and has adequately modelled the load effects for the current and future traffic. The approach and the methods employed to do this have become more elaborate and sophisticated in order to cope with large increases in traffic, a much greater variety of HGVs and more complex bridge designs catering for multiple traffic lanes. Thus the deterministic approach adopted, say, for BS 153: Part 3A [1] using hand calculations and limited data has given way to the probabilistic approach adopted for the long-span loading in the composite version of BS 5400: Part 2 [11].This involved a considerable amount of computing and the collection and processing of large amounts of real data on vehicle weights and lengths.

However, although much work has been done over the years on Type HA loading, very little has been done on the Type HB loading for 'abnormal' vehicles since it was first introduced into a national code, the Joint ICE/IStruct.E *Code of practice* [3], in 1949. The so-called Type C loading in that document later became the Type HB loading in BS 153: Part 3A: 1954 [1] with identical characteristics and configuration. Since then there have been very few changes to the HB vehicle and its application, apart from those arising from metrication. Originally the HB vehicle was of a fixed length but is now specified with a variable length in order to maximise various load effects. The HB vehicle was assumed to be the only vehicle occupying a traffic lane, but now can have HA loading in the same lane albeit with a gap between the HB vehicle and the HA distributed loading.

The HB vehicle with 180 tons on 4 axles was developed at a time when loads of 150–200 tons were the norm and were usually carried on trailers fitted with two bogies. Since then there has been a considerable increase in the numbers of movements of abnormal vehicles, especially those with gross weights up to 150 tons. In addition there are also about 60 movements per year of much heavier vehicle trains which require special authorisation. There are now also quite a wide variety of the heavy load trailers which have more than the 4 axles in service; for example a typical trailer with a capacity of 400 tonnes could have up to 20 axles. As a result of the increased use of abnormal vehicles with multiple axles and the heavier loads being carried, doubts have been expressed from time to time about the adequacy of the HB loading vehicle to match the load effects from real vehicles.

10.2 Regulation of abnormal vehicles

It is important not only to have suitable loading rules for use in design and assessment but also to be able to control the construction and use of the vehicles which are represented by the loading models. This was recognised by the Ministry of Transport who, at the time that the HB vehicle model was being developed, introduced the *Motor vehicles (authorisation of special types) general order* [4] in 1952 to regulate the loading and use of abnormal vehicles weighing up to 150 tons. These regulations have been amended over the years and are currently contained in the *Motor vehicles (authorisation of special types) general order 1979* [5] and its associated (*Amendment*) *order 1987* [6]. It should be noted that 'abnormal' covers all forms of abnormal indivisible loads which can be very wide or very high as well as being very heavy, but for the purposes of this review 'abnormal' will only apply to vehicles which carry heavy loads.

Abnormal vehicles are classified by the regulations into two groups, namely special types general order (STGO) vehicles and special order (SO) vehicles. Special types general order vehicles can have gross weights up to 150 tonnes and limited prior notice of their movements has to be given to the police, and highway and bridge authorities. Special order vehicles are those which are not covered by the STGO regulations and each movement of such vehicles has to be specially authorised by the Department of Transport before it can take place. More details of the two groups are given below.

10.3 STGO vehicles

These vehicles are divided into three categories as follows:

(i) Category 1 consists of vehicles of not more than 46 tonnes gross vehicle weight which are outside the *Construction and use regulations* [7] and which have a total of 5 or more axles.

(ii) Category 2 consists of vehicles of not more than 80 tonnes gross vehicle weight with a total of 5 or more axles and a maximum axle weight of 12.5 tonnes.

(iii) Category 3 consists of vehicles of not more than 150 tonnes with a total of 6 or more axles and a maximum axle weight of 16.5 tonnes.

The allowable maximum loads and axle weights depend upon the distance between the foremost and the rearmost axles and the distances between adjacent axles respectively. The allowable gross weights are for the vehicles actually carrying the load in question and do not include the weights of the towing vehicles. The STGO Category 2 and Category 3 vehicles are subject to speed limits which depend upon the class of road. They are also required to carry a plate showing the initials 'STGO' together with their authorised category, e.g. 'CAT 3'. For most of these vehicles two days' notice of their movement has to be given to the police if the gross weight exceeds 80 tonnes. Two days' notice of movement also has to be given to the appropriate highway and bridge authorities for vehicles less than 80 tonnes, and five days' notice for vehicles heavier than 80 tonnes. Thus for STGO vehicles there is some control over their movements and they may be escorted by the police if their loads are very wide.

10.4 SO vehicles

These are heavier and larger than the STGO vehicles and each movement has to be individually authorised by the Department of Transport under Section 44 of the *Road Traffic Act* [8]. Such vehicles could typically have from 15 to 20 axles and weigh from 300 to 400 tonnes when fully laden. They make much fewer movements than STGO vehicles and are much more closely supervised; a typical journey might well require the movement of street furniture to give room for the vehicles to pass and will have a police escort. The routes to be taken by these vehicles are checked for each journey to ensure that the bridges are strong enough to carry the load, although the necessary information is often already available from records of previous heavy load movements. In some cases it may be necessary to carry out special assessments of any doubtful bridges and strengthen those which are found to be inadequate. Thus, although these vehicles can be several times heavier than the model HB vehicle, their movements are well controlled and it is unlikely that they will cause damage to any structure they are permitted to cross.

10.5 Review of studies of abnormal vehicle loading

In view of the question marks over the effectiveness of the single HB vehicle to model the wide variety of abnormal vehicles actually using the road system, a limited number of studies have been undertaken over the years to investigate the problem. More recently there has been the need to evaluate the abnormal load models proposed for the *Eurocode* for traffic loads on bridges [9]. This section reviews the studies of abnormal vehicle loading in the UK, while the *Eurocode* work is discussed in Chapter 13.

10.5.1 *Highway bridge loading appraisal* [10]

In 1983 Mott, Hay and Anderson carried out a review for the Transport and Road Research Laboratory of the adequacy of the newly adopted BS 5400: Part 2 [11] rules for design loading. As part of the review the consultants looked at the loading from actual abnormal vehicles compared to the loading from the HB vehicle with the variable axle spacing of up to 26 m. A total of 18 vehicles from a DTp list of abnormal vehicles were used in the study, with four of the vehicles being used for more extensive investigations. The gross weights of the vehicles ranged from 267 tonnes on 10 axles to 569 tonnes on 20 axles; each vehicle had two 45.8-tonne tractors at the front and one 45.8-tonne tractor at the rear. The vehicles with their associated HA loading were run over one lane, with no transverse distribution, of a number of idealised bridges representing simply supported and 2-, 3- and 5-span continuous structures with spans from 3 to 200 m. The ratios of the principal load effects from the abnormal vehicles compared to those from the model HB vehicle were determined for support reactions, longitudinal moments and shears at internal supports, and longitudinal moments and shears at a mid-span point. The principal conclusions were as follows:

(i) For very short spans, up to 3 m, the HB vehicle generated a more severe load effect than any abnormal vehicle for all bridge types.
(ii) On the longer spans there was considerable disparity in the load effects from the set of abnormal vehicles compared to those from the HB vehicle.
(iii) For spans greater than 75 m the ratio of the load effects tended to the ratio of the gross weight of the particular abnormal vehicle to the weight of the HB vehicle (i.e. 180 tonnes).
(iv) The variation between abnormal vehicles and the HB vehicle was not consistent for the different load effects.
(v) The choice of spacing between the internal axles of the HB vehicle was critical to determine the maximum loading effect from the HB vehicle.

The overall conclusion was that the HB vehicle did not adequately represent actual vehicles for all spans and circumstances. It should be noted that all the abnormal vehicles examined were over 150 tonnes weight and would therefore have required Special Orders before they could be moved.

As a result of their studies, the consultants proposed two new design vehicles to cover abnormal loads designated ‘Ab 1’ and ‘Ab 2’. Type Ab 1 was intended to represent loads with weights from 90 to 300 tonnes carried on single bogied trailers with from 2 to 8 axles. Type Ab 2 was to represent loads of from 180 to 740 tonnes carried on vehicles with identical twin bogies with from 2 to 10 axles per bogie. All axle spacings were 1.8 m with either 12 m or 18 m spacing between the inner axles of the Ab 2 vehicle. The vehicles were assumed to represent the gross weight of the trailer and its associated tractor, and the selection of the appropriate design vehicle was to be based on the gross weight of the train of actual vehicles being investigated. The new design vehicles were said to produce load effects which were in reasonably close agreement with those from the actual vehicles.

10.5.2 *Long span bridge loading*: Contractor Report 16 [12]

As part of their studies for the derivation of a revised long-span loading for normal vehicles, which is described in Chapter 6, the Flint and Neill Partnership also looked at the influence of the STGO vehicles on traffic loading. From the records over eight years, of the weights and lengths of STGO vehicles supplied by one local authority, it was noted that a large proportion of such vehicles had train weights of between 70 tonnes and 100 tonnes, with only a few trains exceeding 100 tonnes. A histogram of the weight of the abnormal loads divided by (length + 0.9 m) was used to derive a 95 per cent characteristic value of the weight per unit length of the vehicles which was

Table 10.1. Limiting weights of general order vehicle—full HA [12]

Influence line length	100	200	400	800	1600
Single lane effects					
Line 1: Moment	150	105	150+	150+	150+
Line 2: Shear	150+	150+	150+	150+	150+
Line 3: Tower load	140	100	150	150+	150+
Dual carriageway effects					
Line 1: Moment	150+	150+	150+	150+	150+
Line 2: Shear	150+	150+	150+	150+	150+
Line 2: Torsion	150+	150+	150+	150+	150+
Line 3: Tower load	150+	150+	150+	150+	150+

Notes:
Full HA loading accompanying vehicle: load factor = 1.30.
Units: tonnes force, m.

found to be approximately 5 tonnes/m. This figure was taken as the basis for the assumption that the length of any vehicle used in the ensuing study was equal to its weight divided by 5.0.

Three influence line shapes were selected for the studies to represent mid-span moments, end shear and torsion and the axial load in a suspension bridge tower. For a range of influence line lengths from 100 to 1600 m a uniformly distributed load of length $W/5$ m, representing a STGO vehicle of weight W tonnes, replaced the KEL and part of the equivalent UDL for that loaded length. It was assumed that any design to the ULS would be based on the load effects from the larger of either 1.5 times the worst loaded lane UDL plus the KEL, or 1.3 times the UDL and the accompanying 37.5 unit HB vehicle with the 25 m gap at the front and rear of the HB vehicle. The value of the UDL was derived from the proposed revision to the long-span loading. Abnormal vehicles of various weights were tried and the load effects from these were compared with the larger of the design effects described above. From each influence line shape and loaded length it was possible to find the maximum allowable weight of the STGO vehicle which with the associated worst lane Type HA loading would lead to a load factor of not less than 1.3. Results were obtained for both single-lane loading and for a dual 2-lane bridge (shown in Table 10.1).

A second study looked at the value of the associated HA loading to be applied with the HB vehicle. It was accepted that the number of traffic jams in the life of a bridge that are likely to contain abnormal vehicles would be less than the total number of traffic jams. From other data supplied by a local authority it was estimated that the probability that a vehicle at the peak of an influence line would be an abnormal vehicle of over 90 tonnes was approximately 1/35 000. Thus the return period for determining the value of the characteristic distributed load to be combined with an abnormal vehicle of over 90 tonnes was deduced to be 590 h rather than 2400 years. This was seen to justify a reduction in the value of the associated HA load, which was based on a 2400-year return period, when accompanying abnormal vehicles. A histogram derived from the STGO records of vehicle weights, referred to earlier, was used to derive reductions in the HA loading when accompanying

Table 10.2. Limiting weights of general order vehicle—reduced HA [12]

Influence line length	100	200	400	800	1600
Reduction of HA	10%	16%	14%	12%	7%
Single lane effects					
Line 1: Moment	100	100	150+	150+	150+
Line 2: Shear	150+	150+	150+	150+	150+
Line 3: Tower load	95	90	150+	150+	150+
Dual carriageway effects					
Line 1: Moment	150+	150+	150+	150+	150+
Line 2: Shear	150+	150+	150+	150+	150+
Line 2: Torsion	150+	150+	150+	150+	150+
Line 3: Tower load	150+	150+	150+	150+	150+

Notes
HA loading reduced: load factor = 1.50.
Units: tonnes force, m.

STGO vehicles of different weights. These reductions varied for a 100 m loaded length from about 10 per cent for 90 tonne vehicles to 15 per cent for 150 tonne vehicles. The work described above using the three influence lines was repeated but with the reduced accompanying HA distributed load. In this case a load factor of at least 1.5 was deemed to be necessary and again the maximum allowable STGO vehicle weight to satisfy this was determined. The result from the second study is shown in Table 10.2.

The results from both studies indicated that members influenced by 4-lane loading and designed for at least 37.5 units of HB loading provided adequate margins of safety for STGO vehicles up to 150 tonnes under both full and reduced HA loading. For members which are sensitive to single-lane effects with loaded lengths between 100 and 200 m the heavier STGO vehicles would not be adequately catered for by design to 37.5 units of HB loading. A repeat of the study, assuming a tenth of the number of abnormal vehicles, showed that the conclusions were insensitive to the actual number of STGO vehicles.

10.5.3 *Revalidation of traffic loading for abnormal vehicles* [13]

This work, which was undertaken by W. S. Atkins for the Transport and Road Research Laboratory in 1990, consisted of a number of studies and included a comparison of the effects of STGO and SO vehicles with the effects of the current HB loading. This then led to the development of more appropriate loading models for such vehicles. In both cases the studies were limited to loaded lengths of 50 m or less.

The first study was restricted to the effects of loads which could be carried by a single lane. The loads in the other lanes not occupied by the actual abnormal vehicle were taken to be the same as those which would have been applied in design, and so they were not included in the comparisons. A variety of load effects were considered for one- and two-span bridges which were represented as beams; more detailed work was done on 12 m span solid concrete slab decks. Ten STGO vehicles were considered representing the 'worst' from the three categories as well as trailer and tractor combinations. Their gross weights ranged from 80 tonnes to 150 tonnes for the vehicles on their own and to 232 tonnes for a trailer with two tractors. The SO vehicles were selected from those operated by three firms and included 18 trains with gross weights ranging from 264 tonnes to 574 tonnes. An impact factor of 1.8 was assumed to act on one axle for all the STGO vehicles, but no impact was allowed for in the case of the SO vehicles. For the STGO vehicles the HA distributed loading appropriate to the full loaded length was applied (where applicable) up to a distance of 2 m from the front and the rear of the STGO vehicle. In the case of the SO vehicles, all the trains were longer than 50 m and therefore no co-existing HA loading was applied. In the case of the design loading, using the HB vehicle with the code requirement for a 25 m gap at the front and rear of the vehicle it was decided to ignore the effects of any additional UDL which was found to be insignificant for spans up to 50 m. Partial factors of 1.1 and 1.3 were applied to the STGO vehicles (only 1.3 for the HB design loading) to allow for overloading and other unknown effects, but only a partial factor of 1.3 was considered for the SO vehicle cases. The analyses of the two basic bridge types under both the Type HB loading and the abnormal vehicle loading were carried out for different shaped influence lines to generate load effects for bending moments, shears and reactions. The bridges were analysed as beams, but a number of grillage analyses were carried out on the representative concrete slabs to check the validity of the beam approach and to study the effects of transverse bending.

For various reasons it was decided in the case of the STGO vehicles that the recommendations from the work should be based solely on the results using a partial factor of 1.3. The results indicated that the effects of STGO Category 1 vehicles were generally less than the effects from 25 units of HB loading. It was found that while a loading of 30 units of HB was just adequate for all the practical STGO Category 2 vehicles it did not cover the worst of those that were theoretically possible under the regulations; 45 units of HB loading did not cover the effects of all the Category 3 vehicles, especially the tractor/trailer combination. These findings did not specifically cover local effects and were based on vehicles which were more compact than the majority of STGO vehicles in use. For the SO vehicles it was found that very few of the vehicles that were examined were covered by 45 units of HB loading. In some cases, with trailers set at their minimum lengths, effects in excess of 100 units of HB loading were generated.

10.6 Loading models for STGO vehicles

The studies outlined above in 10.5.3 were used as a basis for deriving loading models for STGO vehicles which would fully encompass the effects of all permissible STGO vehicles. The following alternatives were considered:

(i) an increase in the existing HB loading;
(ii) a UDL and KEL with higher intensities than the Type HA loading;
(iii) a new Type HB vehicle with reduced gaps in the associated HA distributed loading.

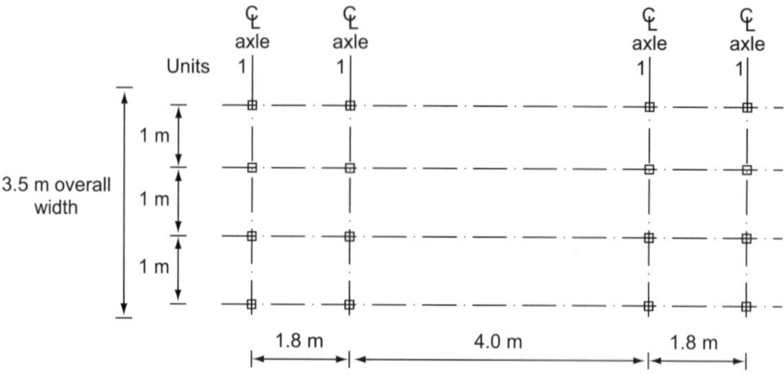

Figure 10.1. Details of proposed replacement for HB vehicle [13]

After evaluating the relative merits of the three alternatives, it was decided that the last option provided the best representation of the global effects of STGO vehicles in traffic. The proposed model vehicle was a shortened version of the existing HB vehicle with a reduced but fixed distance of 4 m between the innermost axles, as shown in Fig. 10.1. This was to be used with the associated HA distributed load on the parts of the lane not occupied by the new HB vehicle, but with a reduced gap of 5 m at the front and rear of the vehicle. In common with the existing HB vehicle, the loading intensity was to be defined in units of 10 kN per axle. Apart from a 150-tonne trailer with two tractors, 45 units of the new HB vehicle covered the effects of all STGO Categories, and 30 units would cover the effects of all Category 2 vehicles. It was also demonstrated that the proposed model could accommodate the co-existence of two STGO vehicles of 120 tonnes and 90 tonnes gross weights respectively on a bridge deck.

10.7 Loading model for SO vehicles

Because there was such a wide variety in the configurations of the 18 vehicles included in the study [13], it was considered that it would be difficult to develop a single model which would model the effects of the different vehicles in a satisfactory manner. It was therefore decided to generate two idealised vehicles, including their tractors, with gross weights of 360 and 720 tonnes respectively. The configurations of trains of intermediate gross weights were to be obtained by interpolation. Details of the vehicles which were devised are shown in Fig. 10.2. For loaded lengths up to 50 m, trailers with different weights but similar bogie lengths produced similar loading effects per unit of gross trailer load which suggested that the tractors contributed relatively little to the overall loading effects for short spans. Therefore including them in the loading models for short spans was unnecessary, but they would still need to be included for the longer loaded lengths.

The solution to the dichotomy outlined above was to base the models solely on the trailers but to apply an amplification factor for the longer loaded lengths where the tractors would contribute to the loading effect. These amplification factors are shown in Table 10.3. It was recommended that in design there should be a 25 m

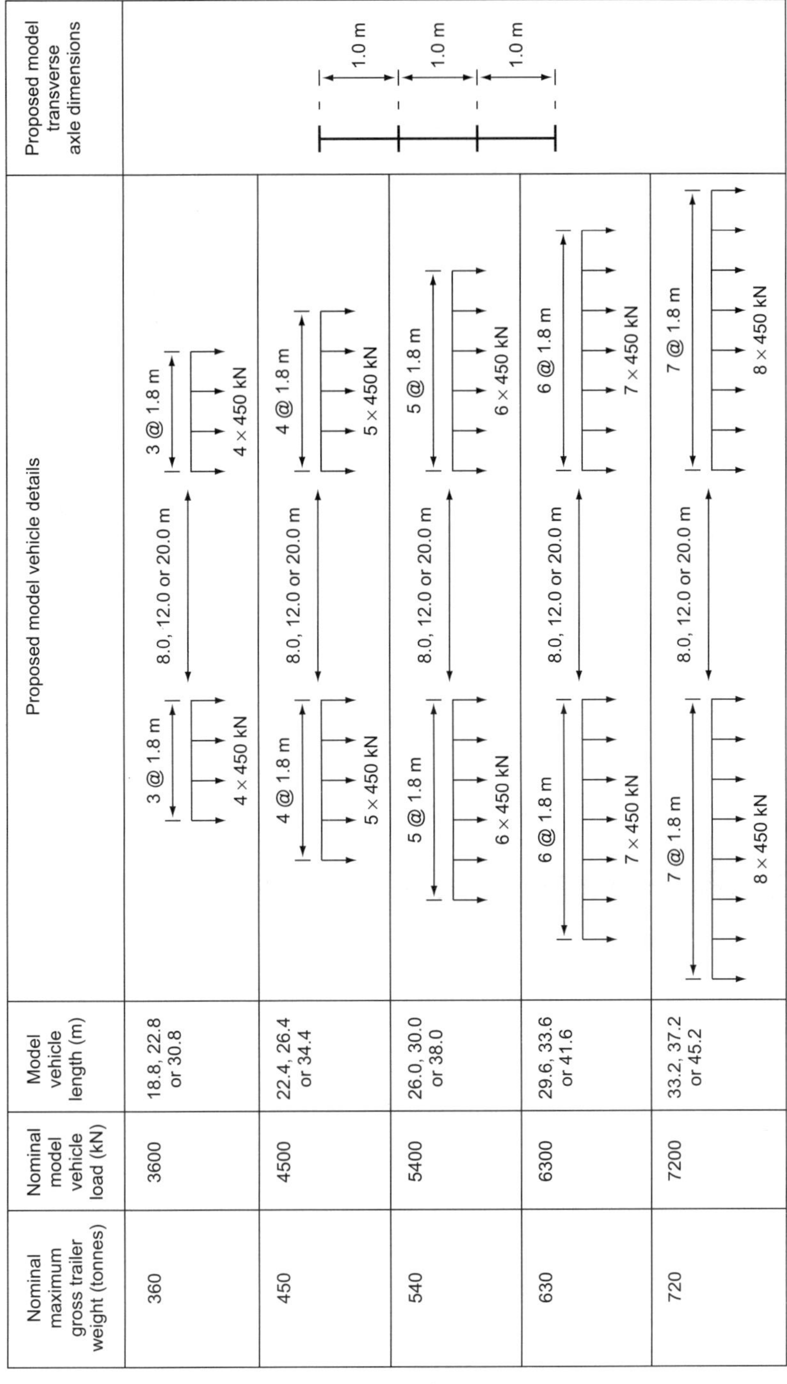

Nominal maximum gross trailer weight (tonnes)	Nominal model vehicle load (kN)	Model vehicle length (m)	Proposed model vehicle details	Proposed model transverse axle dimensions
360	3600	18.8, 22.8 or 30.8	3 @ 1.8 m, 4 × 450 kN; 8.0, 12.0 or 20.0 m; 3 @ 1.8 m, 4 × 450 kN	1.0 m, 1.0 m, 1.0 m
450	4500	22.4, 26.4 or 34.4	4 @ 1.8 m, 5 × 450 kN; 8.0, 12.0 or 20.0 m; 4 @ 1.8 m, 5 × 450 kN	
540	5400	26.0, 30.0 or 38.0	5 @ 1.8 m, 6 × 450 kN; 8.0, 12.0 or 20.0 m; 5 @ 1.8 m, 6 × 450 kN	
630	6300	29.6, 33.6 or 41.6	6 @ 1.8 m, 7 × 450 kN; 8.0, 12.0 or 20.0 m; 6 @ 1.8 m, 7 × 450 kN	
720	7200	33.2, 37.2 or 45.2	7 @ 1.8 m, 8 × 450 kN; 8.0, 12.0 or 20.0 m; 7 @ 1.8 m, 8 × 450 kN	

Figure 10.2. Details of proposed loading models for Special Order vehicles [13]

Table 10.3. Special order vehicle loading model—amplification factors

Loaded length (m)	Amplification factor
2	1.0
30	1.0
40*	1.07*
50*	1.14*
60*	1.21*
70	1.28
1600	1.28

* For $30 < L < 70$. Amplification factor $= 0.007L + 0.79$ where L = loaded length (m).

gap in the associated HA distributed loading at the front and rear of the vehicle. The study also recommended that there should be slight adjustments to the existing traction and braking forces.

10.8 Development of loading model for abnormal vehicle effects

10.8.1 Background

The original version of the bridge assessment code BD 21/84 [14] stated that Type HB loading covering abnormal vehicles need not be applied in the assessment of the bridges covered by the code, and such loading requirements were therefore not included. However, one of the documents [15] giving the technical requirements for the implementation of the DTp's bridge assessment and strengthening programme did require that certain structures should be checked for the number of HB units of loading that they could carry. Later versions of the assessment code have since been amended to say that requirements for HB loading assessment should be sought from the Overseeing Department. The only criteria available at the time were the Type HB loading requirements given in the design code and, as has been described above, these did not accurately represent the load effects induced by real abnormal vehicles. Because the Highways Agency wished to check the ability of all their bridges on motorways and other trunk roads to carry abnormal vehicles as accurately as possible, it was felt that it was essential to develop a new version of the Type HB loading which was tailored for assessment purposes. Accordingly a contract was let by the Highways Agency and the TRL with W. S. Atkins Consultants for the development of such a model. The work was done in collaboration with the Flint and Neill Partnership and checked by Gifford and Partners. The background to the work is described in a report published in 2002 [16] which is summarised below.

10.8.2 Shortcomings of existing HB model

The following shortcomings in the current HB loading model were identified:

(i) The clearance between the abnormal vehicle and the normal traffic in the same lane could be less than the assumed 25 m.

(ii) The effect of impact was not considered because it was assumed that abnormal vehicles travelled at low speeds, whereas in practice they often travelled at speeds over 40 mph with significant impact effect.
(iii) Possible overloading above the authorised weight was not taken into account.
(iv) On routes where there was a high frequency of abnormal vehicle movements there was a good chance that one abnormal vehicle could overtake another or be travelling closely behind in convoy.
(v) The fixed partial load factor provided no flexibility to take account of the effectiveness of the control of movements and weights of abnormal vehicles.
(vi) Compared to actual vehicles the HB loading model could be excessively conservative for very short spans and unconservative for the longer spans.

In order to meet these shortcomings a new loading model, called the Special Vehicle (SV) model, was developed to represent the effects of the abnormal indivisible loads permitted to travel under the *Motor vehicles (authorisation of special types) general order regulations* [5, 6].

10.8.3 Development of SV loading model

In developing the SV loading model the following sources of data on abnormal vehicles were used:

(i) Information on some 15 000 STGO and SO vehicle movements notified to Kent County Council from 1997 to 1999.
(ii) Weigh-in-motion data from the M25 and M4 motorways over a three-month period in the year 2000 from which details of vehicles weighing over 75 tonnes were selected.

The data were carefully screened to remove SO vehicles or vehicles which were considered to be outside the regulatory limits for the STGO categories. In addition, data on military vehicles were collected and a number of hypothetical vehicles conforming to the extreme limits of the STGO regulations were added to the data set. The data were analysed to produce histograms of speeds, gross weights, vehicle lengths, vehicle widths, numbers of axles, maximum axle weights and minimum axle spacings. The data from (i) and (ii) above were not combined but treated independently in the study.

For each vehicle in both data sets the load effects were calculated for influence lines representing mid-span moments and end shear of a simply-supported span and support moments of a continuous beam. Maximum load effects for a range of spans from 2 m to 50 m were determined and compared with the load effects from 45 units of HB loading. These comparisons were done using the nominal HB vehicle with no partial factors or allowances for overloading or impact and omitting any associated HA loading or normal traffic.

Trial SV vehicle configurations were initially chosen to match closely those of the real STGO vehicles of the appropriate categories and their load effects calculated. These configurations were then refined until the load effects from the proposed SV vehicles enveloped the maximum load effects from the relevant category of

Table 10.4. Description of special vehicle models [16]

Vehicle model	Description
SV80	The SV80 vehicle is intended to model the effects of STGO Category 2 vehicles with a maximum gross vehicle weight of 80 tonnes and maximum basic axle load of 12.5 tonnes.
SV100	This is intended to model the effects of STGO Category 3 vehicles with a maximum gross vehicle weight of 100 tonnes and maximum basic axle load of 16.5 tonnes. The data from weigh-in-motion (WIM) and notifications showed that nearly 98 per cent of all Category 3 vehicles had gross weights of less than 100 tonnes and hence a separate model was developed for this group of vehicle. This model also represents mobile cranes within the above gross weight and axle load limits.
SV150	This is intended to model the effects of single STGO Category 3 vehicles with a maximum gross vehicle weight of 150 tonnes and maximum basic axle load of 16.5 tonnes. This model also represents mobile cranes within the above gross weight and axle weight limits.
SV-Train	The SV-Train is intended to model the effects of a single tractor pulling a Category 3 trailer.
SV-TT	The SV-TT models the effects of military tank transporter vehicles with maximum basic axle load of 25 tonnes.

STGO vehicles in the data set. It was found that the finalised SV vehicles provided a much better match to the load effects from the STGO vehicles than the corresponding HB model.

Five SV vehicles were developed to mirror groups of STGO vehicles either on their own or combined with a tractor. Descriptions of the five SV vehicles are shown in Table 10.4 and their axle weights and configurations are shown in Fig. 10.3. The effects of the SV models are generally covered by 45 units of HB loading although in some special cases the SV effects are marginally more onerous. The SV-Train model includes a 46-tonne 3-axle tractor combined with a SV-150 trailer, the tractor being considered to be typical of those used with STGO vehicles. Although the tractor was not representative of the worst configuration used in practice it was felt to be satisfactory in view of the proposal to impose a 180 tonne limit on a tractor plus trailer train in the STGO regulations. The discrete values of the variable spacing between the bogies of the SV vehicles were chosen to minimise any underestimation of load effects compared to the use of a fully variable spacing. With the chosen discrete spacings any shortfall was generally reduced to less than 5 per cent.

10.8.4 Overload factor (OF)

Although there were no reliable data available to determine the level of possible overloading on STGO vehicles, the weight-in-motion (WIM) data indicated that there were a greater number of heavier vehicles than would have been expected from the written notifications. In particular, there were a large number of vehicles

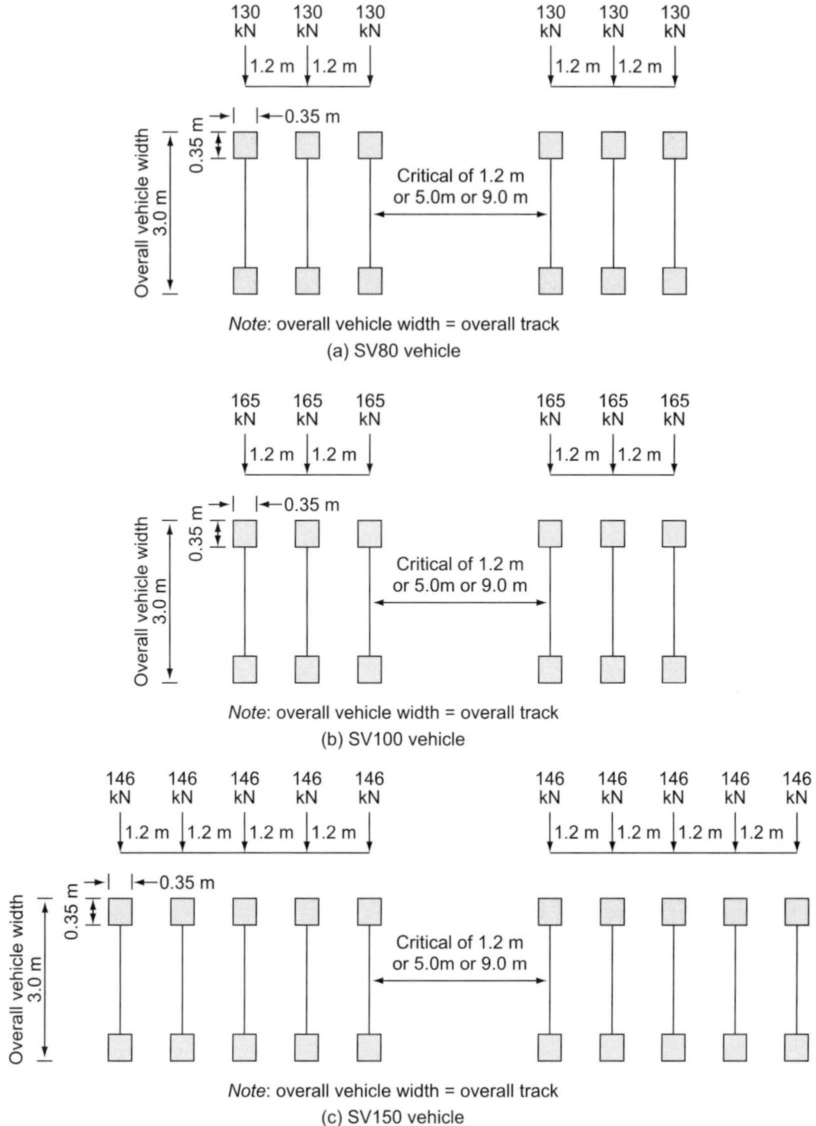

Figure 10.3. Details of proposed special vehicles [16]

with axle weights heavier than the Category limit of 16.5 tonnes and which were not SO vehicles, although this could have been the result of uneven distribution of the trailer load. Based on the above observations an overload factor (OF) was proposed which was 1.2 for the most effective axle and 1.1 for all the others.

10.8.5 Dynamic amplification factor (DAF)

There were no data available on the dynamic effects of abnormal vehicles although it had been observed that the dynamic component of loading was relatively

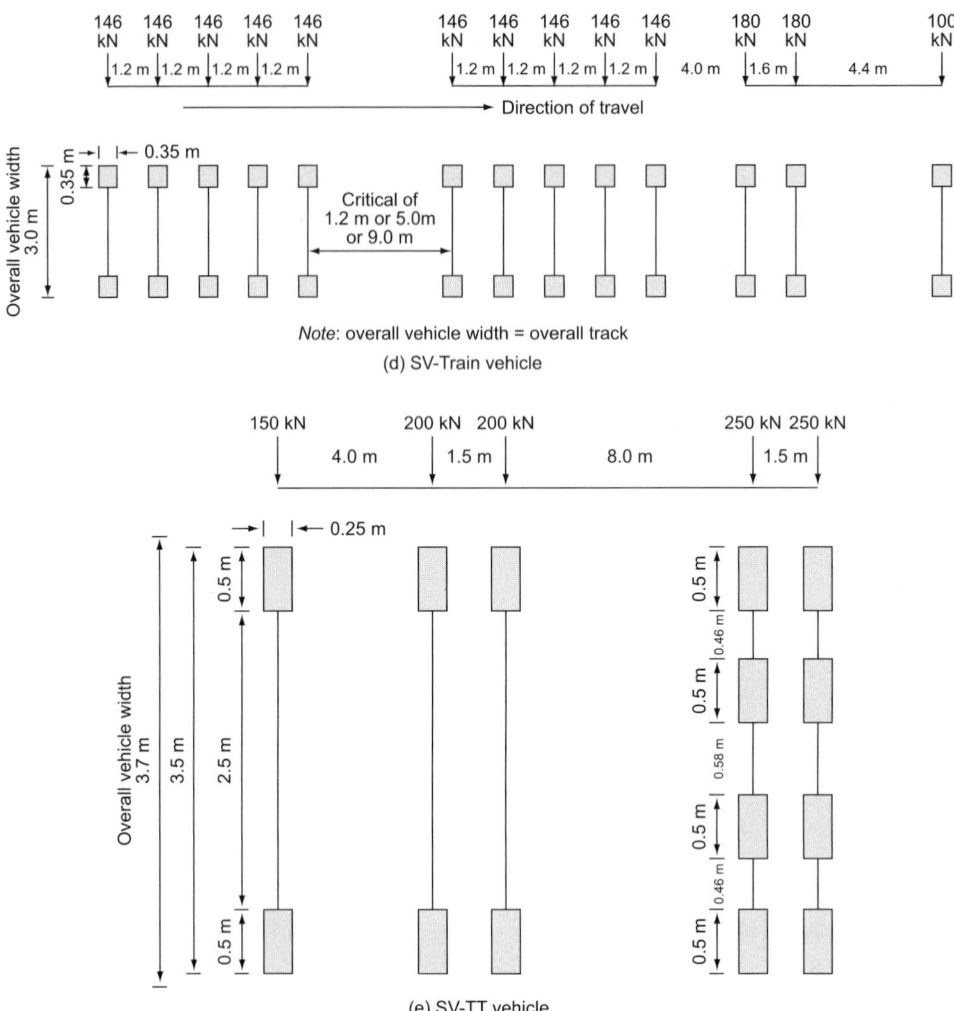

(d) SV-Train vehicle

(e) SV-TT vehicle

Figure 10.3. Continued

independent of the weight of the vehicle. Thus, the increase in loading due to impact could decrease as the vehicle weight increased. A study carried out by Flint and Neill [16] had determined that the characteristic dynamic amplification factor (DAF) for C&U heavy goods vehicles was 1.25 for good road surfaces and 1.38 for poor road surfaces. The STGOs were considered to have lower DAF values than the C&U vehicles because of their heavier weights, lower speeds and generally better suspensions. It was also felt that the impact from the different axles would be uncorrelated and hence the overall dynamic load effect would reduce as the number of axles on the loaded length increased. Where the speed of the STGO vehicle was restricted to less than 10 mph the DAF would reduce to 1.0. Values of the DAF adopted for different axle loads are shown in Figure 10.4.

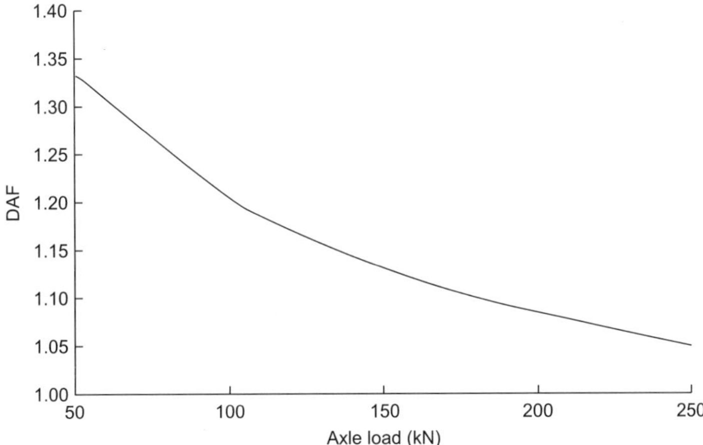

Figure 10.4. Dynamic amplification factors for STGO vehicles as a function of basic axle load [16]. Note: where the speed of the STGO vehicle is restricted to less than 10 mph, the DAF factor is reduced to 1.0

10.8.6 Partial load factors

As it was proposed to deal explicitly with overloading and impact, it was decided to adopt lower partial live load factors than those applied to Type HB loading. Thus the partial factor for use with the SV live load was specified to be 1.1 at the ULS. However, the partial load factor for the associated Type HA loading was retained at the existing value of 1.3.

10.8.7 Masonry arches

Because of the robust nature of the all-terrain fluid suspension and high unsprung axle weights, it was decided that the inertia in the suspension systems of STGO vehicles could give rise to axle lift-off (see Chapter 9). Therefore, in the case of SV vehicles travelling at a normal speed, a lift-off factor of 1.2 was required to be applied to the leading axle of a triple-axle bogie and 0.8 to the trailing axle.

10.9 *Design manual for roads and bridges* BD 86/01 [18]

10.9.1 Background

Based on the work described above, requirements for the assessment of highway bridges and structures for STGO and SO vehicles were issued by the Highways Agency as BD 86/01 [18] in 2001, with colour versions of some of the charts being issued as an amendment in 2002. It was considered that using the new standard would allow the load effects from real STGO and SO vehicles to be determined more accurately and less conservatively than using the HB load model. The standard applies to structures with loaded lengths of less than 50 m and can be applied to all the common materials used in bridge construction. The document deals with STGO vehicles in detail but only gives general principles for dealing with SO vehicles. An innovation in the document is the highlighting of those clauses which are mandatory to distinguish them from other advice and comment. There is frequent reference in the document to the basic assessment code BD 21/01 [19] including requirements for limit states, load combinations, notional lanes and centrifugal force. As much of the contents of the standard have been discussed above in the background to the document, this section will concentrate on the actual application of the document.

10.9.2 Loading

The five models which represent the effects of different categories of STGO vehicles are illustrated with axle weights and configurations, as in Figure 10.3. The requirements for square or rectangular wheel contact areas are shown in each illustration, but for specific vehicles the contact areas notified by the haulier can be used. It should be noted that Type HA loading covers the effects of STGO Category 1 vehicles with a maximum weight of 46 tonnes. Requirements for determining the appropriate values of the overload and dynamic application factors are given, the latter being taken as 1.0 for vehicles travelling at low speeds. The use of a partial load factor of 1.1 on the SV vehicles has already been discussed. Advice is given on the choice of the particular configuration of SV vehicle which is most appropriate for the loaded length being considered.

10.9.3 Application

Diagrams are given which show typical applications of the Type SV vehicles with the associated Type HA distributed loading which is to be determined from BD 21/01. The latter is to be applied to notional lanes across a 2.5 m width but can be replaced by one of the critical vehicles listed in BD 21/01 [19] in certain situations. Only one SV vehicle is to be considered on any one superstructure with the load effects being determined from the maximum of either the vehicle moving at normal speed or the vehicle moving at less than 10 mph. At normal speeds there is to be a space of 25 m at the front and rear of the SV vehicle before any HA loading is applied; at low speeds the gap is reduced to 5.0 m.

10.9.4 Special structures

As in the case of the basic assessment code, the associated loading for certain structures or structures with transverse members is to be in the form of critical vehicles which are to be selected from a list in BD 21/01 [19]. When using one of the alternative methods for masonry arch assessment, the associated HA loading is to be replaced by single, double or triple axles from the list of critical vehicles in BD 21/01.

10.9.5 Reserve factors

The standard requires a so-called 'reserve factor' to be established for the structure concerned for each of the SV vehicles considered. This is defined as the factor on the assessment SV load required to reach first failure and can be calculated with or without the associated HA loading. The 'vehicle rating' of the structure is the most onerous SV vehicle which produces a reserve factor greater than 1.0.

10.9.6 HB-to-SV conversion charts

The standard includes conversion charts which enable the reserve factors for the range of SV vehicles to be calculated from an existing HB rating. In producing the conversion charts the load effects from the SV vehicles have been calculated including the overload factor, dynamic amplification factor and a partial load factor of 1.1, while the load effects for the HB vehicle have included a partial load factor of 1.3. The effects of the associated HA loading have not been included as it was assumed that it would be the same for both SV and HB vehicles. The conversion charts cover six different influence line shapes representing moments, shears and support reactions for both simply-supported and continuous spans.

10.9.7 Management of STGO vehicle movements

The standard includes procedures for the determination of the acceptability of a specific STGO vehicle to pass over a specific structure for which the reserve factors for SV vehicles have already been determined. The assessment begins with a simple screening check where the STGO vehicle characteristics may be compared with corresponding SV vehicles. If it fails this check a more detailed assessment has to be undertaken based on the comparison of the load effects from the STGO vehicle with those from the appropriate SV vehicle using one or more influence line shapes. The procedures allow for reductions in the dynamic amplification factor, overload factor and associated HA loading under certain conditions in order to permit the passage of an STGO vehicle which marginally exceeds the assessed capacity of the structure in question.

References

1. British Standards Institution (1954) BS 153: Part 3A: 1954. *Girder bridges—Part 3: Loads and stresses. Section A: Loads*. London: BSI.
2. Department of Transport (1989) Departmental standard BD 37/88, *Loads for highway bridges*. London: DTp.
3. Joint Committee of Institution of Civil Engineers and Institution of Structural Engineers (1949) *Code of practice for simply supported steel bridges*. London: ICE and IStruct.E.
4. Statutory Instrument 1952 No. 2173 (1952), *The motor vehicles (authorisation of special types) general order 1952*. London: HMSO.
5. Statutory Instrument 1979 No. 1198 (1979), *The motor vehicles (authorisation of special types) general order 1979*. London: HMSO.
6. Statutory Instrument 1987 No. 1327 (1987), *The motor vehicles (authorisation of special types) general order 1987*. London: HMSO.
7. Statutory Instrument 1996 No. 3033 (1996), *The road vehicles (construction and use) (amendment) (No. 7) regulations 1996*. London: HMSO.
8. Road Traffic Act 1988 (c. 52) (1988). London: HMSO.
9. European Committee for Standardisation (1994) ENV 1991-3, *Eurocode 1: Basis of design and actions on structures—Part 3: Traffic loads on bridges*. Brussels: CEN.
10. Mott, Hay and Anderson (1983) *Highway bridge loading and appraisal report*. Unpublished report for Transport and Road Research Laboratory.
11. British Standards Institution (1978) BS 5400: Part 2: 1978 *Steel, concrete and composite bridges—Part 2: Specifications for loads*. London: BSI.
12. Flint and Neill Partnership (1986) *Transport and Road Research Laboratory contractor report 16. Interim design standard: Long span bridge loading*. Crowthorne: TRRL.

13. W. S. Atkins (1990) *The revalidation of traffic loading for abnormal loads and local effects.* Unpublished report for Transport and Road Research Laboratory.
14. Department of Transport (1984) Departmental standard BD 21/84, *The assessment of highway bridges and structures*. London: DTp.
15. Department of Transport (1990) *DMRB*, Vol. 3, Section 4, BD 34/90, *Technical requirements for the strengthening programme for highway structures. Stage 1—Older short span bridges and retaining walls*. London: HMSO.
16. W. S. Atkins Consultants (2002) *Development of a loading model for abnormal vehicle effects.* (Background report to BD 86). Unpublished report for Highways Agency and Transport Research Laboratory.
17. Flint and Neill Partnership (1996) *Probabilistic assessment of short span bridges*. TRL Contractors Report CR16. Unpublished report for Highways Agency.
18. Highways Agency (2001) *DMRB*, Vol. 3, Section 4, Part 19, BD 86/01, *The assessment of highway bridges and structures for the effects of special types general order (STGO) and special order (SO) vehicles*. Norwich: HMSO.
19. Highways Agency (2001) *DMRB*, Vol. 3, Section 4, Part 3, BD 21/01, *The assessment of highway bridges and structures*. Norwich: HMSO.

Chapter 11
Collision loads

11.1 Introduction

In the descriptions of the loading models for the vertical effects of traffic it will have been evident that there have been significant increases in the allowable gross weights of different types of vehicle. However, because of controls on axle spacings and axle weights, the changes in regulations have not, by and large, required matching changes in the values of the loading in the loading models. This is not the case, though, when considering the effects of vehicles colliding with structures, where the gross weight of the vehicle is obviously an important factor. The large increase in the number of HGVs also means that there is now a much greater chance of a bridge support being hit by an errant heavy vehicle. The ability of bridge supports to resist vehicle impacts is, therefore, an important factor to be taken into account in design and assessment. This chapter reviews the development of collision load requirements in the various codes, and gives the derivation of these criteria where possible. The review is limited to vehicle collisions with bridge supports and bridge decks, and does not therefore deal with the loads transmitted to bridge decks as a result of a collision with a parapet.

11.2 Early requirements

Although suggested criteria for collision loads had been in circulation from about 1971, they were not promulgated on a national basis until the issue of BE 5/73 [1] in 1973. This specified a total lateral impact force of 375 kN normal to the traffic and a total force of 225 kN parallel to the traffic. These forces were divided into those at guard-rail level and those at a specified distance above the rail. The values of the forces were derived from impact tests carried out by Jehu and Pearson [2] at the Transport and Road Research Laboratory on different types of safety fence. In the tests, which involved 1.5-ton cars and 5.0-ton coaches, the forces developed on the fences as a result of being hit by the cars travelling at 70 mph were derived from measurements of the decelerations of the vehicles. It appears that the calculated impact forces were scaled up in proportion to the weight of a bus to the weight of a car (5.0 : 1.5) to produce the forces included in BE 5/73.

The collision loads specified in the limit state bridge code in BS 5400: Part 2: 1978 [3] were expressed as nominal loads which were to be multiplied by a partial factor of 1.25 at the ultimate state. This resulted in design forces of 312.5 kN normal to the traffic and 187.5 kN parallel to the traffic which were thus slightly less than the earlier BE 5/73 requirements. This was remedied by the Department of Transport when it implemented BS 5400: Part 2 through Departmental standard BD 14/82 in 1982 [4] when the partial load factor was increased to 1.5. The impact forces for design at the ultimate state then became 375 kN normal to the traffic and 225 kN parallel to the traffic. Also included in BD 14/82 was an additional requirement to consider a single nominal force of 50 kN on the superstructures of bridges with less than 5.7 m headroom. This force was to be applied in the direction of the traffic at any angle between the horizontal and the vertical. These requirements were later incorporated in the major revision of BS 5400: Part 2 which was issued by the Department of Transport in BD 37/88 [5] as the composite version.

11.3 Bridge rehabilitation programme

In the late 1970s and early 1980s a number of problems affecting the safety and durability of trunk-road bridges were identified. This resulted in the formulation of a comprehensive programme of remedial work to tackle the problems, which was announced by the Minister for Roads and Traffic at the end of 1987 [6]. The 15-year bridge rehabilitation programme was later expanded and work began in earnest in 1989. One of the items which was identified was the impact resistance of many bridge supports, which were felt to be inadequate to deal with the heavier goods vehicles now using the roads. An analysis of accident reports which involved HGVs colliding with bridge supports, and which were serious enough to cause an injury to the driver, indicated that there was about a one in ten chance of a trunk-road bridge support being struck during its 120-year lifetime.

Prior to the development of the bridge rehabilitation programme, the Department of Transport had been reviewing the collision load requirements. Back analyses of a number of incidents involving collisions of HGVs with highway structures were carried out to calculate the size of the impact loads generated in the accidents. These calculations were based on the damage sustained by the bridges and the vehicles themselves. All but one of the incidents involved collisions with the superstructures, and the calculated impact forces ranged from 260 kN to 2000 kN with an average of 900 kN. These results led to a recommendation that for supports the total of the two component nominal horizontal loads in the direction of the traffic should be increased to 1500 kN, a tenfold increase, and the total load component at right angles to the traffic to 750 kN. Although large increases were being proposed, these were in line with the requirements in the German standard of the time. The impact load on superstructures was also increased in a *pro rata* manner to 500 kN parallel to the carriageway, but with the addition of considering a load of 250 kN normal to the carriageway. The enhanced criteria were circulated to the Department of Transport Regional Offices in 1990, both for use in new design and for the assessments being carried out under the bridge rehabilitation programme; they were also submitted to the British Standards Institution for consideration as amendments to BS 5400: Part 2 [3].

11.4 Study of heavy goods vehicle collisions with highway structures

Because of the large increases in the collision loading requirements and questions about the applicability of them to all structures, Brian Colquhoun and Partners were asked to investigate the matter and to see whether the requirements should be tailored to fit particular circumstances. Their final report [7] was presented to the Department in 1993. Their conclusions were that the proposed collision criteria were of the right order but that they might be tailored for individual structures by considering the permitted speed at the site and the consequences of any failures.

Besides carrying out comprehensive statistical analyses of the available data on HGV collisions with highway structures, the consultants also looked at the method of determining quasi-static collision loads and at the use of dynamic analysis. For the quasi-static approach they made use of an equation in Appendix A of BS 6779: Part 1 [8], which, for highway parapets, allows the quasi-static force from a collision to be calculated from the vehicle mass and velocity at impact, the location of the centre of gravity and the depth of the vehicle crushing zone. The equation assumes that the vehicle collides with the parapet head-on and slews through 90° on impact before coming to rest. Using the principles in BS 6779: Part 1 [8] it was calculated that a 40-tonne vehicle travelling at 50 mph would generate a collision load of 1500 kN. Confidence in the basic assumptions in BS 6779 was reinforced by the good agreement obtained between calculated decelerations and those measured during collision tests on concrete bridge parapets.

Using the BS 6779 [8] equation, the consultant produced a table of nominal collision loads on bridge supports for different vehicle speeds as shown in Table 11.1. The 'other areas' columns refer to inter-urban roads where there are safety fences,

Table 11.1. Nominal collision load on structure supports [7]

	Urban and speed restricted areas				Other areas	
	Permitted highway speed (mph)				Permitted highway speed (mph)	
	30	40	50	60	60	70
Nominal collision load parallel to carriageways (kN)						
Main	400	700	1000	1500	700	1000
Residual	200	350	500	750	350	500
Nominal collision load normal to carriageway (kN)						
Main	200	400	500	800	400	500
Residual	100	200	250	400	200	250

Table 11.2. ULS collision load on decks [7]

	Permitted speed (mph)				
	30	40	50	60	70
Horizontal ULS load (kN)	250	400	600	800	1050
Vertical ULS load (upwards) (kN)	200	250	400	500	650

soft verges and central reserves that could contribute to the slowing down of a vehicle prior to any collision. This slowing down was estimated to be about 20 mph, and so it was assumed that any collision on a 70 mph road would in fact take place at 50 mph; for urban roads it was assumed that the collision would occur at the speed limit of the road in question. The consultants also produced a table of the horizontal and vertical loads to be applied to a bridge superstructure for different road speed limits as shown in Table 11.2. The horizontal loads were based on calculations for a collision involving a 10-tonne unladen tipper lorry with its back accidentally raised. The vertical loads in the table were based on estimates.

The use of dynamic analysis was also investigated, but it was found that because of the large number of variables involved and the difficulty of estimating their values, the accuracy of the results was questionable. It was considered that, until sufficient research had been carried out, the design of bridge supports for collision loads should continue to be based on the application of quasi-static loads. The conclusions of the study were that collisions which caused significant damage were comparatively rare events and the proposed enhanced impact loads were of the right order.

11.5 *Design manual for roads and bridges* BD 48/93 [9]

The enhanced collision loading on bridge supports was formally promulgated through the *Design manual for roads and bridges* as Departmental standard BD 48/93 in 1993. The standard was intended to be used for the assessment and possible strengthening of any suspect bridge supports which had been identified under the bridge rehabilitation programme. The loading criteria for supports were the same as previously circulated, namely a total of 1500 kN parallel to the traffic and 750 kN at right angles to the traffic with a partial factor of 1.5. The requirements superseded those in BD 37/88 [5] but retained the exemption for supports more than 4.5 m from the edge of the carriageway. In a slight change of presentation, reference to the loads on the supports being transmitted from the safety fence was omitted, and the text just referred to the 'main load component' and the 'residual load component'. Another change from BD 37/88 was that the main load component parallel to the traffic was now greater than the main load component at right angles to the carriageway and also higher than the residual load component parallel to the traffic.

Departmental standard BD 48/93 allows the nominal collision loads to be reduced when using a quasi-static method of analysis. The reduction factor $30/(30 + m)$, where m is the mass of the support member in tonnes, is based upon momentum conservation and assumes that the support member alone participates in the dynamic response to the collision.

Further reductions are allowed for the assessment of other elements of the structure such as foundations and deck slabs directly connected to support members.

11.6 *Design manual for roads and bridges* BD 60/94 [10]

The enhanced collision loading requirements for design were promulgated through the *Design manual for roads and bridges* as Departmental standard BD 60/94 in 1994. The requirements for supports were the same as those in BD 48/93 [9] and superseded those in BD 37/88 [5], but any reduction in the load values was limited to the treatment of foundations. Departmental standard BD 60/94 also contained enhanced requirements for the collision loads to be applied to the superstructures of bridges over highways. A single-point load of 50 kN was replaced by a nominal load of 500 kN parallel to the carriageway and a nominal load of 250 kN normal to the carriageway. These superstructure loads only were to be applied if the headroom at the structure concerned was less than 5.7 m.

11.7 Computer simulation of heavy goods vehicles collisions with bridges

In order to get a better understanding of the behaviour of bridges after being struck by HGVs, the Department of Transport (later to be the Highways Agency) commissioned Ove Arup and Partners to investigate the problem using computer simulation techniques. The development of the computer models and the results obtained were described in a report by Arup in 1994 [11]. In the study over 150 cases of actual collisions involving a variety of HGVs were analysed using simple computer models.

To calibrate three more complex computer models, a number of full-scale tests were undertaken at the Motor Industry Research Association's testing ground [12]. Three impact tests were carried out involving vehicles being run into a specially constructed precast prestressed concrete beam bridge which included a single reinforced concrete column. The tests used the following types of heavy goods vehicle:

(i) A water filled 4-axle rigid tanker weighing 30 tonnes colliding with the reinforced concrete column at 40 mph.

(ii) An empty 4-axle rigid tipper lorry weighing 11 tonnes with its body raised colliding with the test bridge superstructure at mid-span at 30 mph.

(iii) A tracked excavator weighing 12 tonnes carried on a low loader and tractor weighing 13 tonnes and travelling at 40 mph; the arm of the excavator was raised so that the elbow joint between the front and main arm hit the bottom flange of the edge concrete beam in the test bridge deck.

The results of the tests and details of the test vehicles and the bridge were reported to the Highways Agency in 1995 [13].

The findings from the computer analyses and comparisons with the test results, together with implications for design, were also reported to the Highways

Agency in 1995 [14]. The comparisons may be summarised as follows:

(i) Tanker to column impact: reasonable agreement in velocities of tanker between computer predictions and test measurements. The peak impact force calculated from decelerations was 5600 kN with an average over time of 2300 kN.
(ii) Raised tipper to deck impact: reasonable agreement between calculated velocities and test measurements during initial period after impact. The calculated peak horizontal force was 900 kN with an average of 360 kN. The calculated peak vertical force was 780 kN with an average of 140 kN.
(iii) Excavator to deck impact: good agreement between computer predictions and test measurements for displacements and velocities after impact. The calculated peak horizontal force was 1700 kN with an average of 460 kN. The calculated peak vertical force was 630 kN with an average of 210 kN.

The consultants examined the findings from the project to assess the implications regarding the rules for collision loads on bridges. In the case of the tanker/column collision it was felt that the enhanced design forces, and the equation in BS 6779: Part 1 [8], gave good estimates of the average forces experienced by the column during the test. However, it was considered that the design forces considerably underestimated the peak forces and would need to be at least doubled. For the bridge deck impacts involving the raised tipper and the excavator arm the impact forces generated during the collisions were considered not to be indicative of the maximum forces that could be achieved because of premature failures of the test vehicles themselves. A computer simulation of the excavator test was therefore rerun with the vehicle moving at a higher speed, and this yielded average horizontal and vertical forces which on the superstructure were in-line with the enhanced values included in BD 60/94 [10], although the peak values were considerably higher. The general conclusion was that the then current design rules for collision loads were inadequate for both bridge supports and superstructures, although it was noted that in the case of the latter the damage to the bridge deck was insignificant.

Despite the recommendations in the Arup report [14], the collision loads in BD 60/94 [10] have not so far been amended. In a recent paper in the *Proceedings of the Institution of Civil Engineers, Structures and Buildings* [15], Das, of the Highways Agency, wrote the following comment on the Arup studies:

> One conclusion was that although the instantaneous impact forces were of the order of 7 MN, the new impact design requirement of 1.5 MN was reasonable as an average for static design.

Perhaps this conclusion was influenced by the fact that although accidental collisions occur quite frequently, so far only one trunk-road bridge has been completely dislodged by a collision. Also, as mentioned, the damage to the superstructures in the full-scale tests was insignificant and the column failed in a brittle manner before its full moment capacity was reached. There were, therefore, questions about the full-scale test results which may have raised doubts about using the limited information to introduce significant changes to the design rules.

References

1. Department of the Environment (1973) Technical Memorandum (bridges) BE 5/73, *Standard highway loadings*. London: DoE.
2. Jehu, V. J. and Pearson, L. C. (1972) Transport and Road Research Laboratory report LR 502, *Vehicle impact tests on the tensioned-beam and open box crash barriers*. Crowthorne: TRRL.
3. British Standards Institution (1978) BS 5400: Part 2: 1978, *Steel, concrete and composite bridges—Part 2: Specification for loads*. London: BSI.
4. Department of Transport (1982) Departmental standard BD 14/82, *Loads for highway bridges—Use of BS 5400: Part 2: 1978*. London: DTp.
5. Department of Transport (1989) Departmental standard BD 37/88, *Loads for highway bridges*. London: DTp.
6. Department of Transport (1987) Press notice No. 581, *Bridge rehabilitation programme*. London: DTp.
7. Brian Colquhoun and Partners (1993) *Collision of heavy goods vehicles with highway structures: Vol. 1 Report, Vol. 2 Appendices*. Unpublished report for Department of Transport.
8. British Standards Institution (1998) BS 6779: Part 1: 1998, *Specification for vehicle containment parapets for metal construction*. London: BSI.
9. Department of Transport (1993) *DMRB*, Vol. 3, Section 4, Part 7, BD 48/93, *The assessment and strengthening of highway bridge supports*. London: DTp.
10. Highways Agency (1994) *DMRB*, Vol. 1, Section 3, Part 5, BD 60/94, *The design of highway bridges for vehicle collision loads*. London: HMSO.
11. Ove Arup and Partners (1994) Report No. AT94-09, *Computer simulation of HGV collisions with bridges—Phase 1*. Unpublished report for Highways Agency.
12. Motor Industry Research Association (1994) Reports Nos. 436024, 436025 and 436033, *A series of full scale dynamic impact tests on a bridge structure by various types of heavy goods vehicles*. Unpublished reports for Highways Agency.
13. Ove Arup and Partners (1995) Report No. AT 95/06, *HGV/bridge collision test reports*. Unpublished report for Highways Agency.
14. Ove Arup and Partners (1995) Report No. AT 95/01, *Computer simulation of heavy goods vehicle collisions with bridges—Phase 2*. Unpublished report for Highways Agency.
15. Das, P. C. (2001) Assessment loading criteria for bridge decks, piers and parapets. *Proceedings of Institution of Civil Engineers, Structures and Buildings*, **146**(4), pp. 411–421. London: Thomas Telford.

Chapter 12
Control of traffic loading

12.1 Introduction

The previous chapters have looked at the development of the traffic loading criteria for the design and assessment of highway structures. It will have been noted that the criteria have been amended and enhanced from time to time to ensure that they have continued to provide adequate margins of safety over the effects produced by the traffic of the day. The major changes, particularly for longer spans, have been mainly due to increases in the numbers of vehicles rather than due to any increases in vehicle or axle weights. While the numbers of vehicles on the roads cannot easily be controlled, it has been the practice in the UK to control the construction of all types of vehicle which are used on public roads and to regulate their use. It has also been recognised that there is a very important relationship between the design of vehicles and the loading criteria used for the design of highway bridges. Changes to the former should not be made without considering the effects on the bridges themselves if the loading criteria are to remain valid for any length of time.

This chapter reviews the regulations which govern the construction and use of motor vehicles in the UK, and describes how proposed new vehicle configurations are assessed. It deals mainly with 'normal' vehicles as the regulations for 'abnormal' vehicles have already been discussed in Chapter 10. The chapter also looks at other means for controlling traffic loading such as by alterations to road layouts. Finally the derivation and enforcement of the present system of weight limits on weak structures is considered.

12.2 Vehicle regulations

A review of some of the Acts of Parliament which regulated the construction and use of the early locomotives and motor cars is given as an appendix to the Report of the Joint Committee on Loads on Highway Bridges in 1918 [1]. The Acts date back to

1861 and it appears that it was the effect of wheel loads on the road surface itself that was the important consideration, rather than the loading on bridges. Thus, in the Locomotives Act, 1861, the allowable axle loading depended upon the width of the tyre, with a maximum axle load of 12 tons being allowed on wheels with 12 in wide tyres. The Locomotives Act, 1898, allowed county and borough councils to give permission for loads of up to 16 tons on wheels with tyres not less than 8 in wide. The *Heavy Motor Car Order*, 1904, stated that the axle weight of a loaded heavy motor car should not exceed 8 tons with the gross weight of the vehicle not to exceed 12 tons. Incidentally there was a maximum speed limit of 8 mph for such vehicles.

Since the days described above, when steam locomotives and 2-axle motor vehicles were the norm, there has been a tremendous change in the development and use of road vehicles. The first *Motor vehicles (construction and use) regulations* were introduced in the 1920s to control the gross weights and axle weights and spacings of vehicles. These regulations continue to this day, although now referred to as the *Road vehicles (construction and use) regulations*, and are in the process of being replaced by the *Road vehicles (authorised weight) regulations*. Apart from giving the allowable axle and gross weights for all types of road vehicle, the regulations also deal with many other aspects of vehicle design such as their width, braking, steering, lighting and signalling, etc.

As described by Ford in a paper to an International Association for Bridge and Structural Engineering (IABSE) colloquium in 1975 [2], the *C&U regulations* since the 1920s have always had regard to the bridge design loadings of the day when deriving allowable vehicle configurations. But before the days of electronic computation there was some difficulty in fully comparing the effects of proposed new vehicles with the formula-type loading to ensure that they were covered by the design loading. As a consequence, the controls on vehicle construction imposed by the *C&U regulations* tended to be on the conservative side.

About the time that the first bridge assessment code BE 4 [3] was being prepared, a computer progam was written by the MoT for calculating the bending moments and shears produced in a simply supported span by trains of actual vehicles. A description of the program, OPTAX, is included in Appendix 1. The program, which could include an impact factor on the worst axle, was used to produce 'control' values of bending moments and shears for various spans by enveloping the effects produced by trains of vehicles which were already permitted under the *C&U regulations*. Any proposed new vehicle with specific axle weights and spacings was run through the program, and the bending moments and shears for different spans were then compared with the equivalent control values. It was thus fairly easy to decide whether the vehicle configuration was acceptable, and if not what changes might need to be made to the axle weights or spacings. It was felt that it was better to compare vehicles directly in this way on a single-lane basis, rather than to make comparisons with the formula loading used for design, because the latter contained adjustments to take account of different forms of construction.

Following consultation with industry and based on the approach described above a very comprehensive set of *C&U regulations* was published in 1973 [4]. These gave the allowable axle weights and spacings and maximum gross weights for rigid and

Table 12.1. C&U vehicle and axle weights

a. Rigid vehicles

No. of axles	*Gross vehicle weight (tonnes)*
2	17.00
3	24.39
4	30.49

b. Articulated vehicles

No. of axles		
Tractor	*Trailer*	*Gross vehicle weight (tonnes)*
2	1	24.39
2	2	32.52
2	3 or more	38.00
3	1	32.52
3	2 or more	38.00

c. Single axle

Gross axle weight (tonnes)
10.5

d. Bogies

No. of axles	*Overall minimum axle spread (m)*	*Gross bogie weight (tonnes)*
2	1.02	16.26
2	1.85	20.34
3	1.4	18.00
3	2.7	22.5
3	2.6+	24.0+

Notes:
The maximum gross vehicle and axle weights allowable under the C&U regulations are tabulated. In the case of vehicles these are associated with specified minimum axle spacings. Full details of these spacings and the corresponding gross vehicle weights for closer spacings are given in the regulations.
The C&U regulations permit 3 axled bogies of 24 tonnes gross weight, provided that they have an air or fluid suspension. These bogies should be considered when undertaking arch assessments.

articulated vehicles and also the rules for the closely spaced axles of 2- and 3-axle bogies. These regulations were said to provide a more realistic relationship between axle weights and spacings and bridge design than had previously existed in the UK. They ensured that the effects of possible trains of the heaviest legally loaded vehicles were always slightly less than those derived from the design loading on spans of from 2 m to 30 m. Since then there have been various issues of the regulations and associated amendments which have allowed heavier vehicles and slightly increased axle loads, but all changes will have been checked to ensure that there has been little or no significant exceedence of the control values. Thus the introduction of heavier vehicles has usually been accompanied by requirements for the loads to be carried on a greater number of axles. The maximum gross vehicle and axle weights allowable under the latest *C&U regulations* [5, 6] are shown in Table 12.1. Details in the changes to the allowable gross vehicle and axle weights since 1904 are shown in Table 12.2.

Table 12.2. Maximum allowable axle and gross vehicle weights

Year	Axle weight	Rigid vehicles			Articulated vehicles			
		2 axles	3 axles	4 axles	3 axles	4 axles	5 axles	6 axles
1904[(1)]	8 T	12 T	12 T	12 T				
1927	8 T[(2)]		19 T					
1930				22 T				
1931					19 T	19 T		
1941	8 T[(3)]							
1955	9 T	14 T	20 T	24 T	20 T	24 T		
1964[(4)]	10 T	16 T	22 T	28 T	22 T	32 T		
1972[(5)]			24 T	30 T	24 T			
1973[(6)]	10.17 t	16.26 t	24.39 t	30.49 t	24.39 t	32.52 t		
1983	10.5 t[(7)]						38 t	
1988	10.5 t[(8)]	17 t						
1993	10.5 t[(9)]		26 t[(10)]	32 t[(10)]	26 t[(10)]	35 t[(10)]		
1999	11.5 t[(11)] 10 t[(13)]	18 t	26 t	32 t	26 t	38 t	40 t 44 t[(14)]	41 t[(12)] 44 t[(14)]

Notes:
T = tons, t = tonnes.
[(1)] No. of axles not mentioned. 3- and 4-axle vehicles may not have existed. [(2)] 7.5 T for 3 or more axles.
[(3)] For all axles. [(4)] First limits based on axle spacing (tandem axle 18 T). [(5)] Tandem axle 20 T.
[(6)] Metrication. [(7)] Only on sole driving axle if gross vehicle weight is more than 32.52 t.
[(8)] On driving axle of all 2-axle rigids and all articulated tractors. [(9)] On sole driving axle of any vehicle.
[(10)] With air suspension or equivalent. [(11)] Drive axle. [(12)] 10.5 t drive axle. [(13)] Non-drive axle.
[(14)] Inter-modal journeys.

12.3 European vehicles

It is not always possible to control the maximum weights of vehicles and axles when there are pressures for change which originate outside the country. Thus, as a result of European legislation, the UK has had to accept increases in axle weights and gross vehicle weights for vehicles used in international trade. The UK, however, obtained a derogation which meant that the more onerous vehicles would not have to be accepted in the UK until the end of 1998. The details of the vehicles were originally promulgated by the European Commission under Directive 85/3/EEC with various amendments with the final set of amendments being promulgated in Directive 91/60/EEC. As described in Chapter 5, the need to cater for the more onerous European vehicles led to adjustments having to be made to the loading curve for short spans. The most significant changes were the adoption of an 11.5-tonne drive axle and a gross weight of 40 tonnes.

The European vehicles have now been incorporated in the UK regulations under *The road vehicles (authorised weight) regulations* 1998 [7], although *The road vehicles (construction and use) regulations* 1986 [5] as amended still apply to existing vehicles for the time being. Details of the maximum axle and gross vehicle weights under the 1998 regulations are shown in Table 12.3. It will be noticed that these give some discretion to vehicle manufacturers by providing formulae for determining the maximum authorised weights based on the distances between axles.

Table 12.3. AW vehicle and axle weights [9]

a. Rigid vehicles

No. of axles	*Gross vehicle weight (tonnes)*
2	18.00
3	26.00
4	32.00

b. Articulated vehicles

No. of axles		
Tractor	*Trailer*	*Gross vehicle weight (tonnes)*
2	1	26.00
2	2	38.00
2	3 or more	40.00
3	1	36.00
3	2 or more	40.00 and 44.00[(2)]
3	3	41.00[(1)]
3	3	44.00[(1)]
3	(articulated bus)	28.00

c. Single axle

	Gross axle weight (tonnes)
Driving axle	11.5
Non-driving axle	10.0

d. Weight by reference to axle spacing—rigid vehicles

The maximum authorised weight in kilogrammes in the table below is the distance between the centres of outer axles of the vehicles (in metres) multiplied by the factor in the third column and rounded up to the nearest 10 kg, if that number is less than the maximum authorised weight.

Description of vehicle	*Number of axles*	*Factor to determine maximum authorised weight*	*Maximum authorised weight (kg)*
Rigid motor vehicle	2	6,000	18,000
Tractor unit	2	6,000	18,000
Trailer which is not a semi-trailer or centre-axle trailer	2	6,000	18,000
Rigid motor vehicle	3	5,500	25,000, 26,000[(3)]
Tractor unit	3 or more	6,000	25,000, 26,000[(3)]
Trailer which is not a semi-trailer or centre-axle trailer	3 or more	5,000	24,000
Rigid motor vehicle	4 or more	5,000	30,00, 32,000[(3)]
Articulated bus	Any number	5,000	28,000

e. Weight by referene to axle spacing—articulated vehicles

The maximum authorised weight in kilogrammes for a articulated vehicle in the table below is the distace between the kingpin and the centre of the rearmost axle of the semi-trailer (in metres) multiplied by the factor in the third column ad rounded up to the nearest 10 kg, if that number is less than the maximum authorised weight.

Description of vehicle combination	*Number of axles*	*Factor to determine maximum authorised weight*
Articulated vehicle	3 or more	5,500

The maximum gross vehicle and axle weights allowable under the *Road vehicles (authorised weight) regulations 1998* are tabulated. In the case of vehicles these are associated with specified minimum axle spacings. Full details of these spacings and the corresponding gross vehicle weights for closer spacings are given in the regulations. For existing vehicles and for 44-tonne intermodal transport journeys, the *Road vehicles (construction and use) regulations 1986* as amended still apply.

Notes:

(1) 10.5 tonne axle. (2) International intermodal transport journeys only (permitted under the Road Vehicles (Construction and Use) Regulations 1986 as amended). (3) The driving axle if it is not a steering axle is fitted with twin tyres and road-friendly suspension, or each driving axle has twin tyres and no axle weight exceeding 9,500 kg.

12.4 Traffic loading restrictions

Although the traffic loading on bridges can be controlled to some extent by the use of regulations such as those described above, it is also possible to limit traffic loading by introducing physical restraints on the carriageway. Thus, the imposition of lane restrictions, such as by installing kerbs or barriers, can reduce the number of lanes of traffic on a weak bridge. In extreme cases the number of lanes can be limited to just one lane by the use of signal controlled one-way working. In the case of some long-span bridges which carry a high proportion of HGVs, the drivers of such vehicles have been encouraged to leave a good distance between themselves and the next heavy vehicle in front by white lines painted at intervals across the carriageway. The idea is that there should be only one HGV at any time in the space between successive white lines, thus reducing the density of the traffic.

More recently on some long bridges where there are concerns about the bridges becoming overloaded, use is being made of WIM devices to monitor the total weight of traffic on the bridge at any instant or to check for overloaded vehicles. On the Tamar Bridge, which has recently been widened, WIM pads have been installed in the road surface at the approach to and at the exit from the bridge span. These weigh the vehicles as they move onto the bridge and as they leave the bridge and thus enable a running total of the weight of vehicles on the bridge at any one time to be calculated. At the moment this information is just being monitored, but it helps the bridge managers to decide if the bridge is becoming overloaded and whether some form of manual intervention may be required to reduce the traffic flow.

12.5 Weight restrictions

During the production of the bridge assessment code BD 21/84 [8] there was considerable discussion about the use and effectiveness of weight restrictions for controlling the weights of vehicles on weak bridges. At that time the weight restriction signs were applicable either to single-axle weights or to the weight of the vehicle at the time it was crossing the bridge concerned. These restrictions were introduced at a time when the majority of vehicles had only 2 axles and there was usually a simple relationship between the gross weight of the vehicle and the weight of the heaviest axle. However, the introduction of multiple-axle bogies and the increasing number of axles on vehicles meant that the axle and vehicle weight restrictions were no longer relevant. Thus a bridge might be able to carry 20 tonnes spread over 4 axles but would be overloaded if the same load was to be carried on 2 axles. Similarly an axle weight limit of 8 tonnes might be acceptable on a 2-axle vehicle with a gross weight of 12 tonnes, but would be unacceptable on a 4-axle rigid vehicle which could weigh up to 24 tonnes. A major disadvantage was that operators of overweight vehicles could only be prosecuted if the weight of the vehicle at the time could be shown to exceed the signed limit. This meant that vehicles had to be weighed at an authorised weighbridge before any action could be taken, and it was not always possible to find a convenient one to do this.

It was therefore decided that the weight restrictions should be updated so that they could cater for all the various types of vehicle in use as well as being easy to enforce. In deriving a simpler and more practical system it was realised that there would have to be a fairly crude approach and that vehicles which, for example, were only part-laden might be unfairly penalised. The key to the new system was that there were to be only a limited number of weight restriction levels based on the maximum allowable gross weights of certain easily recognised types of vehicle. The maximum allowable weights of heavy vehicles were readily available as they were recorded on special metal plates attached to each vehicle. Thus the 26-tonne limit in the latest version of the assessment code BD 21/01 [9] is the maximum weight for 3-axle vehicles, and the 18-tonne limit the maximum for 2-axle vehicles. The 7.5-tonne limit is the limiting weight beyond which the driver requires an HGV driving licence; vehicles heavier than 7.5 tonnes also have special light reflecting red and yellow markings at the rear. The 3-tonne limit is the maximum for cars and so no goods vehicles, apart from light vans, should be crossing a bridge with such a weight restriction. Details of the size and markings of weight restriction signs are contained in *The Traffic Signs Regulations and General Directions 1994* [10]. The updated system makes it possible for an enforcement officer to be able to tell at a glance whether a particular vehicle should be crossing a weak bridge or not. When weight restrictions are applied to a bridge, the confirming Traffic Order will state whether there are any exemptions such as for buses or for empty vehicles returning to their depot. In addition to the signing of weight restrictions on weak bridges, Trading Standards Officers have powers to stop vehicles in roadside checks to see whether they or any of their axles are overloaded. Vehicles which are suspected of not complying with the *C&U regulations* are either weighed on the spot with portable weighing equipment, or are escorted to a nearby weighbridge. In some areas these are permanently installed in specially constructed lay-bys.

References

1. Report of Joint Committee of Concrete Institute, Institution of Municipal and County Engineers and Institution of Municipal Engineers (1918) *Loads on highway bridges*. London: The Concrete Institute.
2. Ford, J. (1975) *Derivation of regulations for vehicle weights and axle spacing from the design loading for short span bridges*, Paper 11, IABSE (British group) Colloquium on highway bridge loading, Cambridge.
3. Ministry of Transport (1967) Technical Memorandum (bridges) No. BE 4, *The assessment of highway bridges for construction and use vehicles*. London: MoT.
4. Statutory Instrument 1973 (1973) *The motor vehicles (construction and use) regulations 1973* as amended. London: HMSO.
5. Statutory Instrument 1986 No. 1078 (1986), *The road vehicles (construction and use) regulations 1986*. London: HMSO.
6. Statutory Instrument 1995 No. 3051 (1995), *The road vehicles (construction and use) (amendment) (No. 6) regulations 1995*. London: HMSO.
7. Statutory Instrument 1998 No. 3111 (1998), *The road vehicles (authorised weight) regulations 1998*. London: HMSO.

8. Department of Transport (1984) Departmental standard BD 21/84, *The assessment of highway bridges and structures*. London: DTp.
9. Highways Agency (2001) *DMRB*, Vol. 3, Section 4, Part 3, BD 21/01, *The assessment of highway bridges and structures*. Norwich: HMSO.
10. Statutory Instrument 1994 No. 1519 (1994), *The traffic signs regulations and general directions 1994*. London: HMSO.

Chapter 13
Eurocodes

13.1 Background

In 1975 the Commission of the European Communities embarked on a programme based on Article 95 of the Treaty of Rome to harmonise technical specifications. This included establishing a set of harmonised technical rules for the structural design of construction works. These *Eurocodes* for the design and construction of building and civil engineering works were to eventually replace the differing rules in member states. There are ten structural *Eurocodes* covering the technical aspects of structural and fire design of buildings and civil engineering works, each one consisting of several parts. Traffic loads on bridges, EN 1991-2, is a part of EN 1991—*Eurocode 1: Actions on structures*. Details of all the *Eurocodes* and the expected release dates are given in a Special Issue of *Civil Engineering* in the *Proceedings of the Institution of Civil Engineers* [1].

At present most *Eurocodes* have been released in a 'pre-standard' form, designated EuroNorm Vornorms or ENVs which will be converted to full EuroNorm (EN) status after a period of trial use and the receipt of comments from users. Currently each ENV may be used for design in conjunction with a 'National Application Document' (NAD) for the member state where the structure is located. These application documents allow the country concerned to impose its own national safety levels and requirements for serviceability and durability, as well as authorising the use of national supporting standards where necessary. In the final EN versions, National Annexes will replace the NADs and allow national preferences to be specified where left open in the *Eurocode*. As soon as a final draft of each *Eurocode* is available a period of transition will begin, the last part of which will be a co-existence with national codes and standards. At the end of the period national standards will be withdrawn, provided all the *Eurocode* parts which are relevant to the particular form of design are available. The pre-standard version of *Traffic loads on bridges* is contained in ENV 1991-3 and has been issued together with the UK NAD as a draft for development [2] by the British Standards Institution. At present the final version of the EN is being prepared together with the UK *National Annexe*.

13.2 Development of ENV 1991-3 [2]

Work on the development of a *Eurocode* for the loads on highway bridges started at the end of 1987 under the auspices of the Commission of the European Communities. A steering committee composed of the representatives of the national authorities of several member states met to decide upon the scope and contents of the proposed *Eurocode*. Several sub-groups were set up to look at particular topics and to carry out detailed studies. At the end of 1991 responsibility for the work was transferred to the Comité Européen de Normalisation (CEN) and given to two project teams, PT6 for road traffic loads on highway bridges and PT7 for loads on railway bridges. All the sub-groups had submitted their reports by then and their results and recommendations were drafted into the pre-standard ENV 1991-3 which was published in 1994.

13.3 ENV 1991-3 [2], scope and contents

Pre-standard ENV 1991-3 covers the imposed loads on road and railway bridges together with footbridges, but this chapter will only consider the traffic loading for road bridges. For road bridges the code is applicable to bridges with individual spans ranging from 5 m to 200 m and with carriageway widths not greater than 42 m. There are four models for determining the main vertical loads from traffic representing the different types of traffic or different design situations. In addition there are rules for determining secondary live loads such as those due to acceleration and braking and centrifugal effects. The requirements for accidental loading cover collisions with bridge supports and the effects of errant vehicles on areas such as footways and cycle tracks.

13.3.1 Load Model 1

This is the main traffic loading system and consists of concentrated and uniformly distributed loads and is intended to cover the global and local effects of normal traffic. The concentrated load model (known as the tandem axle system) consists of a pair of axles with four identical wheels having a square contact area of 0.4 m × 0.4 m. For spans greater than 10 m the pair of axles can be replaced by a single axle of the same total weight. The UDL system has a constant value irrespective of the span length, although there are different values for the individual traffic lanes. The distributed load is to be applied to the unfavourable parts of the influence surface. The basic values of the tandem axle system and the UDL system, which include dynamic amplification, are shown in Table 13.1 for the different notional lanes. These can be multiplied by 'adjustment factors' to allow for site specific situations, but in most cases the factors are to be taken as unity. Details of Load Model 1 are shown in Figure 13.1.

13.3.2 Load Model 2

This model consists of a single-axle load, including dynamic effects, which can be applied to certain areas as shown in Figure 13.2. It is to cover the effects of

Table 13.1. Basic values for Load Model 1 [2]

Location	Tandem system Axle loads Q_{ik} (kN)	UDL system q_{ik} (or q_{rk}) (kN/m^2)
Lane number 1	300	9
Lane number 2	200	2.5
Lane number 3	100	2.5
Other lanes	0	2.5
Remaining area (q_{rk})	0	2.5

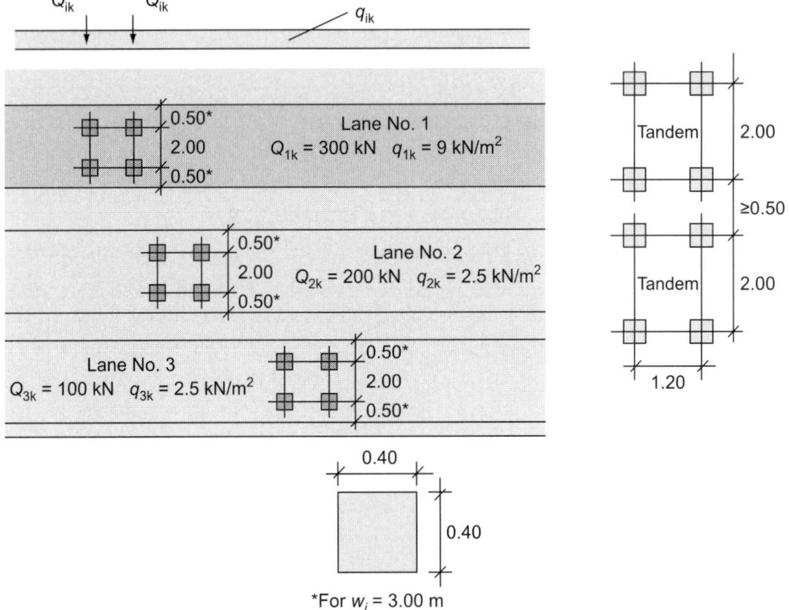

Figure 13.1. Load Model 1 [2]

normal traffic on very short structural elements and is to be used on its own to determine purely local effects. The basic axle load is 400 kN and it can be modified by adjustment factors if necessary.

13.3.3 Load Model 3

This consists of sets of axle loads at given spacings to represent special (abnormal) vehicles carrying heavy loads on designated abnormal load routes. The weights of the listed SVs range from 600 to 3600 kN with from 4 axles to 18 axles, and with axle loads of 150, 200 or 240 kN depending on the class of the vehicle concerned. This model is intended to be used only as requested by the client and is suitable for both global and local effects. These vehicles are intended to generate equivalent load effects for the range of actual vehicles in general use which do not comply with the national regulations governing the maximum weights and dimensions of normal vehicles. The SVs are assumed to move at very low speeds and therefore contain no

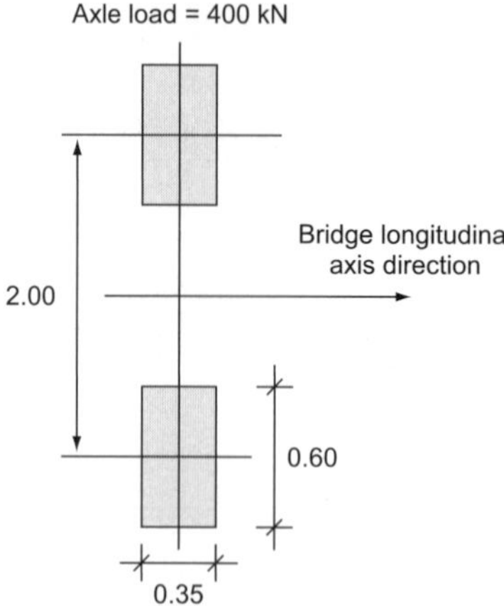

Figure 13.2. Load Model 2 [2]

allowance for dynamic effects. Details of the special vehicles are given in Table 13.2; the axle spacing is 1.5 m.

The selected SVs are applied to either one or two adjacent lanes depending on the weight and dimensions of the vehicle concerned. Each notional lane and any remaining area is to be loaded by the main loading system as well, but using frequent rather than characteristic values for the intensities of the concentrated and distributed loads. On the lanes actually occupied by the SV the main loading system is not applied for a distance of 25 m in front of and behind the SV. Figure 13.3 shows the application of the SV and the associated loads.

Table 13.2. Classes of special vehicles [2]

Total weight	Composition	Notation
600 kN	4 axle-lines of 150 kN	600/150
900 kN	6 axle-lines of 150 kN	900/150
1200 kN	8 axle-lines of 150 kN	1200/150
	or 6 axle-lines of 200 kN	1200/200
1500 kN	10 axle-lines of 150 kN	1500/150
	or 7 axle-lines of 200 kN + 1 axle-line of 100 kN	1500/200
1800 kN	12 axle-lines of 150 kN	1800/150
	or 9 axle-lines of 200 kN	1800/200
2400 kN	12 axle-lines of 200 kN	2400/200
	or 10 axle-lines of 240 kN	2400/240
	or 6 axle-lines of 200 kN (spacing 12 m) + 6 axle-lines of 200 kN	2400/200/200
3000 kN	15 axle-lines of 200 kN	3000/200
	or 12 axle-lines of 240 kN + 1 axle-line of 120 kN	3000/240
	or 8 axle-lines of 200 kN (spacing 12 m) + 7 axle-lines of 200 kN	3000/200/200
3600 kN	18 axle-lines of 200 kN	3600/200
	or 15 axle-lines of 240 kN	3600/240
	or 9 axle-lines of 200 kN (spacing 12 m) + 9 axle-lines of 200 kN	3600/200/200

Note: Vehicle widths of 3.00 m for the 150 and 200 kN axle-lines, and of 4.50 m for the 240 kN axle-lines are assumed.

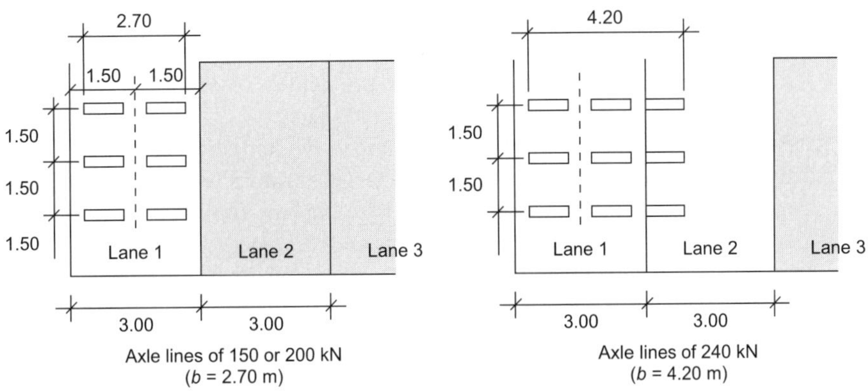

Figure 13.3. Location of the special vehicles on bridge deck [2]

13.3.4 Load Model 4

This represents crowd loading and consists of a UDL of 5 kN/m^2 and is only to be applied when requested by the client.

13.4 Derivation of Load Models 1 and 2

13.4.1 Background

The background to the development of Load Model 1 (LM1), the main load model, was described in a paper presented at an IABSE conference in 1996 [3]. At the start of the work, traffic data from a number of countries were analysed by one of the working groups. Some of the information had only been recorded on the inside, or slow, lanes, but nevertheless all the data contained useful information about axle weights, axle spacing, distance between vehicles and the length of vehicles. The durations of the various data collections ranged from a few hours to 800 h. Although there was quite a large scatter in the measurements of axle weights and total vehicle weights, it was found that there was reasonable agreement in the maximum values for a mean return period of one day between the different locations. It was found that the traffic parameters from the different countries were not very different, especially when comparing the maximum daily values of axle and vehicle weights. However, rather than base a traffic load model on a mix of all the traffic data it was decided to use one set of data recorded on the A6 motorway in France, a 2-lane dual carriageway near the city of Auxerre. Because of the number of international vehicles using the A6, it was felt that the traffic data would give a good representation of European traffic as a whole. It was assumed that the real traffic records included an inherent dynamic effect with a magnification factor of 1.10.

13.4.2 Methodology

The procedure for the development of the main load model LM1 consisted of the following main steps:

(i) Determination of target values of various actual traffic load effects which were to be reproduced by the design load models. These load effects were to be extrapolated to correspond to a probability of exceedence of 5 per cent in 50 years or a return period of 1000 years.

(ii) An investigation to find and define the load model which was best able to reproduce the target values for loaded lengths from 5 to 200 m. The determination of the target values, largely based on the traffic data recorded at Auxerre, took into account different extrapolation methods, traffic composition, different influence lines for the various load effects and the dynamic effects from flowing traffic.

13.4.3 Extrapolation methods

Three traffic parameters were considered in the studies, namely extreme axle and lorry weights, extreme total loads on a span, and extreme load effects. Various methods of extrapolation were used in all the different cases but the results were generally found to be in good agreement with each other. The methods used and the results obtained are summarised below; more detailed descriptions are given in a document [4] which contains the reports of some of the working groups involved in the preparation of the load models.

In the case of axle loads and lorry loads, three methods of extrapolation were used based on fitting standard distribution curves to the histograms of the axle and lorry weights recorded over a week. The extrapolated values obtained from the three methods were generally quite similar although there were differences of up to 20 per cent in some cases. Using the results it was proposed that the characteristic values of the loads at the ULS should be as follows:

- single axle—between 250 and 300 kN
- double axle—between 380 and 400 kN
- triple axle—between 490 and 530 kN
- lorry weight—between 780 and 900 kN (excludes authorised special vehicles).

A second set of analyses were undertaken to predict the maximum of the total load (or the equivalent UDL) on a given lane length based on the traffic recorded in the slow lane at Auxerre. The following three traffic situations were considered:

(i) natural free-running traffic using the recorded inter-vehicle spacings;

(ii) congested traffic including cars with a constant inter-vehicle spacing between extreme axles of between 2 and 5 m;

(iii) congested traffic with lorries only and a constant inter-vehicle spacing of between 2 and 5 m.

In all, five statistical approaches were used to predict the maximum total loads for different return periods. These included analytical methods using mathematical models to describe the distributions of vehicle weights with specific lengths along the bridge, or simulation techniques using Monte-Carlo methods to create randomly composed convoys of vehicles. In the latter case the vehicles and their characteristics were chosen randomly from a garage of vehicles whose content mirrored the make-up of the recorded traffic. The total weights of the vehicles generated were

Table 13.3. Maximum total load (kN/m) for 1000-year return period [4]

Traffic state	Span (m)			
	20	50	100	200
Free-flowing	46.2	31.1	21.2	13.8
Congested with cars	49.4	40.4	34.3	27.7
Congested—no cars	51.1	42.9	37.8	34.2

Note: Figures obtained from averaging results of different methods of extrapolation.

Table 13.4. Equivalent distributed load (kN/m) for 1000-year return period [4]

Traffic state	Span (m)			
	20	50	100	200
Free-flowing	60.34	34.26	22.76	17.70
Congested with light vehicles	51.42	40.45	35.70	31.33
Congested without light vehicles	52.87	42.40	36.50	33.63

Note: Based on maximum mid-span bending moment of simply supported beam

extrapolated to various return periods using Gaussian, Poisson or extremal distributions. One approach involved the use of a program to generate traffic jams which had initiation points at different locations along the bridge span. By and large, there was reasonable agreement between all the approaches when comparing the maximum total loads for different spans and different return periods, although the jam simulation gave consistently higher results than the other methods. The average of the extrapolated load densities for a single lane for a 1000-year return period for free-flowing and congested traffic are shown in Table 13.3.

A third set of analyses looked at the prediction of extreme load effects for a limited number of bridge influence lines. The extrapolation methods applied were generally similar to those used to predict the total loads although they mainly involved simulation techniques. Both free-flowing and congested traffic were considered. It was found that for a 1000-year return period the equivalent UDLs predicted agreed well for most spans, although those produced by the extremal method using simulated jams were significantly higher. Table 13.4 shows the 1000-year return period equivalent UDLs derived for various spans and for different traffic situations.

The general conclusion from the trials of different methods of extrapolation was that none of the methods was too inaccurate to be used and that the results obtained were not very sensitive to the adopted extrapolation technique. Two of the methods were later used in developing the target values for calibrating the proposed loading models.

13.4.4 Dynamic effects

The dynamic behaviour of the vehicles and bridges was examined using numerical simulations based on assumed roughness values for the carriageway surface. For spans greater than 15 m an average roughness was taken; for spans less than 15 m

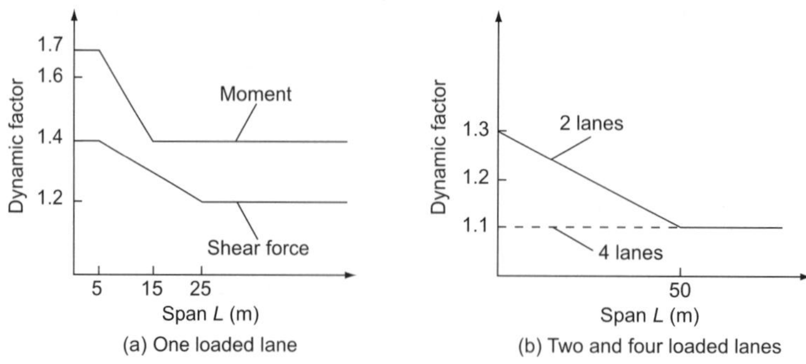

Figure 13.4. Average global dynamic factors [4]

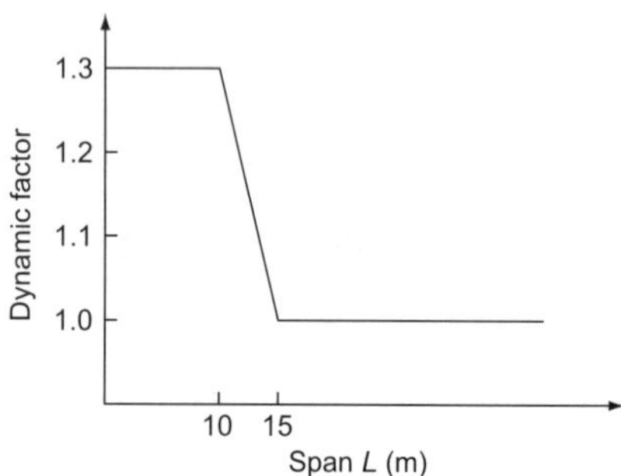

Figure 13.5. Average local dynamic factors [4]

the roughness was represented by a 30 mm thick plank. Figures 13.4 and 13.5 show average values of the global and local dynamic factors which were used to determine the load effects from the recorded traffic data. These dynamic functions were combined statistically with the static effects in recognition of the fact that the maximum dynamic effect did not necessarily occur at the same time as the maximum static effect. The combined effects were used to generate the target values for calibrating the proposed load models.

13.4.5 Loading patterns

In order to determine the target values for calibration it was necessary to define the loading patterns or scenarios to be considered, including the composition of the traffic occupying each lane. For the purpose of the exercise the calibrations were done for a dual 2-lane carriageway. It was also necessary to consider both free-flowing and congested traffic. As the target values were determined by several different working groups there were inevitably some differences in the assumptions made, but they can be generally summarised as follows:

(i) Flowing traffic
Lane 1 (slow)—extrapolated traffic loads based on data from A6 motorway slow lane with 25 per cent lorries.

Lane 2 (fast)—daily maximum of A6 slow-lane recorded traffic (no extrapolation), or recorded traffic in A6 fast lane with 10 per cent lorries.
Lanes 3 and 4 (other carriageway)—generally daily mean loads recorded on A6 fast lane, or with limited percentage of lorries.

(ii) Congested traffic
Lane 1 (slow)—based on A6 slow-lane traffic without cars.
Lane 2 (fast)—based on extrapolated A6 slow-lane flowing traffic, or simulated jam situation with 5 m spacing between vehicles.
Lanes 3 and 4 (other carriageway)—daily maximum and daily mean of A6 slow-lane traffic respectively.

All the results from the different groups were used to determine a single set of target values, with the target values for very short spans being adjusted to take proper account of local effects.

13.4.6 Influence lines

Nine different influence lines were used to calculate the loading effects from the different traffic patterns and to calibrate the main loading models. The shapes of the influence lines are shown in Table 13.5. For each influence line nine span lengths were considered, namely 5, 10, 20, 30, 50, 75, 100, 150 and 200 m. Of the nine influence lines those for the mid-span bending moment of a simply supported beam (I1) and for the central support of a two-span continuous beam (I9) were found to be the most significant for the calibration of the load model.

Table 13.5 Influence lines considered

No.	Representation	Description of the influence line
I1		Maximum bending moment at mid-span of a simply supported beam.
I2		Maximum bending moment at mid-span of a double-fixed beam with an inertia that varies strongly between mid-span and the ends.
I3		Maximum bending moment on support of the former double-fixed beam.
I4		Minimum shear force at mid-span of a simply supported beam.
I5		Maximum shear force at mid-span of a simply supported beam.
I6		Total load.
I7		Minimum bending moment at mid-span of the first of the two spans of a continuous beam (the second span only is needed).
I8		Maximum bending moment at mid-span of the first span of the former continuous beam.
I9		Bending moment on central support of the former continuous beam.

13.4.7 Calibration of Load Model 1

The aim of the calibration studies was to develop a loading model which would allow for dynamic magnification effects, would cover the different traffic patterns or scenarios including free flowing and congested traffic, and would cover both global and local effects. It was to consist of both concentrated and distributed loads with the minimum distributed load being set at $2.5\,kN/m^2$ based on existing national standards. Operational research techniques were used to home in on the optimum load model by minimising the differences between the load model effects and the target values.

The calibration of LM1 was performed by the successive consideration of a single loaded lane, two loaded lanes and finally four loaded lanes. Generally it was found that in a longitudinal direction the best fitting model consisted of both concentrated and distributed loads. For global effects a single concentrated load in each lane would have been satisfactory but, to model local effects realistically, two concentrated loads were necessary. It was also found that for the best accuracy the intensity of the distributed load should reduce with increasing loaded lengths, but to simplify the application of the loading it was agreed to have a constant distributed load even though this meant some loss of accuracy.

Further studies on loaded lengths less than 5 m led to the discovery of the need to increase the intensity of the concentrated loads on the first and second lanes with a reduction in the intensity of the associated distributed load. The distance between the axles of the tandem-axle system was increased from 1.0 m to 1.2 m and it was found that it was not necessary to apply the system to the fourth lane or above. The final version of Load Model 1, as incorporated in ENV 1991-3, is shown in Figure 13.1.

The basic values of the concentrated and distributed loads may be modified by adjustment factors which allow reduced loading to be adopted in specific situations. These could be applied, for instance, for a particular class of road where the volume and mix of traffic is significantly different from that used in deriving the load models.

13.4.8 Load Model 2

Calculations showed that the tandem-axle system in LM1 did not adequately cover all the local effects for all vehicles. Therefore for local effects, in particular those on orthotropic slabs, it was agreed to make good any deficiencies by the introduction of Load Model 2 (LM2), consisting of a single axle with a total load of 400 kN, as shown in Figure 13.2. This load model also allows different wheel contact areas to be taken into account.

13.5 Application of loading

13.5.1 Lane widths

The carriageway is divided into an integral number of notional lanes which are a standard 3 m wide; any width left over is known as the ‘remaining area’. The appropriate concentrated and distributed loads are applied to the notional lanes which are independent of any traffic lanes painted on the road surface. Only a distributed load is applied to the remaining area. For carriageways less than 5.4 m wide only one notional lane is taken, with the rest of the carriageway being loaded with the standard 2.5 kN/m^2. The location and the numbering of the notional lanes is determined in such a way that the effects from the load models produce the most adverse effects. It should be noted that there are no explicit lane factors as variations in lane loading are taken into account by specifying different intensities of loading for each particular numbered notional lane.

13.5.2 Partial live load factors

The partial factors on actions (loads), together with the combination and reduction factors which apply to combinations of traffic and other types of loading and for loading situations which are more frequent than the characteristic loading, are given in an appendix to the *Eurocode*. It is intended that this information should eventually be transferred to the EN 1990 *Eurocode—Basis of structural design* [5], which is the lead document for the whole set of structural *Eurocodes*. The partial factor on traffic loads at the ULS for persistent and transient situations is 1.35, but this reduces to 1.0 for accidental loading situations. For the serviceability limit state the partial factor on traffic loads in persistent and transient situations is 1.0; accidental situations are not considered at the serviceability limit state.

13.5.3 Load combinations

Where traffic loads are to be combined with other forms of loading the intensities of loading are reduced by means of the reduction factor ψ_0 which is given in a table in Appendix C to the code. Frequent rather than characteristic values of the loading are to be taken in these situations to reflect the lower probabilities of different forms of loading reaching their maxima at the same instant.

13.5.4 Reduction factors

Other reduction factors are included in Appendix C which deal with infrequent, frequent and quasi-permanent loads, the last being rarely used. For instance when special vehicle loading is applied the associated main traffic loading from LM1 is reduced from a characteristic value to an infrequent value by the application of the appropriate reduction factor. The use of the reduction factors will be discussed in more detail later on when considering the National Application Document.

13.6 Other types of loading

13.6.1 Accidental loads

These include requirements for vehicle collisions with bridge supports and decks, the effects of heavy wheel loads on unprotected footways and cycle tracks, and the effects of vehicle collisions with parapets. In the first case the support should be designed to resist a force of 1000 kN in the direction of travel and a force of 500 kN at right angles to the direction of travel, both forces being applied at a height of 1.25 m above road level. In the case of bridge decks no specific requirements in terms of loads are given. In the case of accidental footway loading one axle of the LM1 tandem-axle system is to be taken with a 200 kN axle weight and

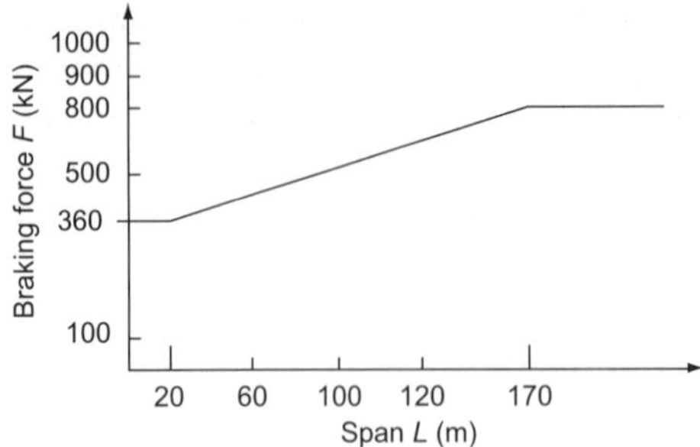

Figure 13.6. Braking force [4]

it can be applied either parallel to or at right angles to the direction of the traffic. No other variable load is to be included. The horizontal force to be transferred to the deck as a result of a collision with a parapet is to be taken as 100 kN. This is to be applied in conjunction with a vertical traffic force of 150 kN. For deformable barriers the collision force is to be taken from the results of the full-scale approval tests on the barriers.

13.6.2 Braking

The characteristic braking force, which has the same magnitude as the acceleration force, is to be taken as a longitudinal force at carriageway level. It is calculated as a fraction of the total maximum vertical traffic load applied to Lane 1 but is not to be greater than 800 kN. It may be applied along the axis of any lane and may be considered as uniformly distributed over the loaded length.

The magnitude of the braking force was derived from a study carried out in Germany based on a number of assumptions, which were confirmed by tests carried out in Switzerland. It was assumed that there was a set of identical lorries with uniform spacing crossing a bridge at the same speed. The reaction time for braking between consecutive vehicles was taken as the spacing between vehicles divided by the speed of the convoy. The braking force exerted by a lorry was taken as being proportional to its weight with a factor varying from 0.6 to 1.0 depending upon the type of lorry and its weight. Various simulations with different parameters were carried out which led to the relationship of braking force being described as a function of span length as shown in Figure 13.6. The formulae given in ENV 1991-3 [2] were derived directly from this relationship.

13.6.3 Centrifugal force

The characteristic centrifugal force is to be taken as a transverse force acting at finished carriageway level. For a radius of curvature of the carriageway less than 200 m the centrifugal force is 0.2 times the total maximum weight of the concentrated loads of the tandem system from LM1. For a radius of curvature greater than 200 m and less than 1500 m the centrifugal force is $40/r$ times the total maximum weight of the concentrated loads, where r is the radius of curvature in metres. If the radius of curvature is greater than 1500 m the centrifugal force may be ignored.

The requirements for dealing with centrifugal force, Q_{tk}, were derived from the equation $Q_{tk} = V^2 \times Q_v/g \times r$ where V is the vehicle speed (m/s), Q_v is the corresponding vertical force (kN), $g = 9.81\,m/s^2$ and r is the radius of curvature (m). The values of Q_{tk} given in ENV 1991-3 [2] correspond to a vehicle speed of about 70 km/h. The speed was chosen as being representative of the speed of heavy vehicles which have the major influence on centrifugal effects.

References

1. *Eurocodes* (2001) *Proceedings of the Institution of Civil Engineers, Civil Engineering*, **144**, Special Issue 2, November.
2. British Standards Institution (2000) DD ENV 1991-3: 2000, *Eurocode 1: Basis of design and actions on structures—Part 3: Traffic loads on bridges*. London: BSI.
3. Bruls, A., Calgaro, J. A., Mathieu, H. and Prat, M. (1996) IABSE report 74, ENV 1991—Part 3: *The main models of traffic loads on road bridges*. Background studies. IABSE conference on basis of design and actions on structures, Delft.
4. Calgaro, J. A. (1997) ENV 1991-3: Traffic loads on bridges, *Calibration of the main loading system—Background studies*. Unpublished report, Paris.
5. Comité Européen de Normalisation (2001) pr EN 1990—*Eurocode: Basis of structural design—Stage 34*, April 2001, CEN.

Chapter 14 National Application Document (NAD)

14.1 Background

14.1.1 National requirements

From the start of the development of the *Eurocodes* it was recognised that there was a clear distinction between design codes and national requirements, especially those concerned with safety. While *Eurocodes* could provide a common set of rules for the design and construction of structures it would still be the responsibility of each country to set the levels of safety and reliability to be achieved by the use of the codes in the particular country. Consequently the ENVs contain certain parameters whose values are printed in brackets ([ ]) and whose values may be adjusted by each country to ensure that their particular national requirements can be satisfied. These alternative national values are to be included in a National Application Document (NAD) which, for each relevant ENV *Eurocode*, describes how the *Eurocode* is to be applied for the design of structures in the particular member state. Eventually, after experience has been gained in the use of the ENVs, they will be reviewed and converted into fully fledged ENs. The NADs with the bracketed values will then be replaced by National annexes attached to each *Eurocode*. The NAD for use with ENV 1991-3: 1995 for designing road bridges in the UK was published by the British Standards Institution in 2000 together with the English version of the ENV [1]. The development of the NAD is described in this chapter.

14.1.2 Calibration studies

Almost from the start of the work on the *Eurocode* for traffic loading the Flint and Neill Partnership were involved through a contract with the TRL in providing support for the UK contribution to the work. Once the final version of ENV 1991-3 had been published Flint and Neill were asked by the Highways Agency to assist in the development of the UK's NAD including the 'boxed values' relevant to highway loading. The work included the derivation of suitable values for the characteristic load adjustment factors α which can take account of site specific conditions and different lane loading, and the derivation of the reduction factors ψ which take account of combined loads and also loading events which occur more frequently than characteristic events. The scope of the work also included the review of abnormal vehicle loading and fatigue loading.

Initial comparative studies [2] were based on a statistical analysis of UK traffic including data collected on the M6 at Warrington in 1990. However, the Highways Agency subsequently had a change of policy and decided that the main highway loading requirements in ENV 1991-3 should be calibrated against the current UK standard BD 37/88 [3]. The results from this later work were then incorporated into the NAD which was issued by BSI. This chapter covers both the initial study and the subsequent study which will be referred to as the 'final' study. The report on the latter [4] also provides a comprehensive background to the various clauses in the NAD which refer to highway loading.

14.2 Initial parametric calibrations

14.2.1 Scope

The initial parametric calibrations of the UK requirements against the boxed *Eurocode* values for the main traffic loading models, LM1 and LM2, were carried out for a dual carriageway with two lanes in each direction. Three influence lines were considered for both congested and flowing traffic. The influence lines represented the maximum mid-span bending moment, the simple end shear for a simply-supported beam, and the moment at the central support of a two-span continuous beam. The analyses were carried out for various spans ranging from 5 to 200 m. Congested traffic conditions were considered for all spans, but flowing traffic was only considered for spans up to 50 m as congested traffic governed the loading beyond that point.

14.2.2 Congested traffic

For congested traffic the overall state load effects were derived using a statistical vehicle/axle model derived from weigh-in-motion data recorded on the M6 at Warrington. The histograms of load effects for each influence line and span were used to derive the characteristics of a Type 1 extremal distribution of maximum effects. From the distribution the effects for different return periods were obtained, namely, for a 1000-year period, for a one-year period and for a return period of one week. Dynamic effects were only included when considering single-lane loading, it being assumed that the measured data included an impact factor of 10 per cent. The traffic conditions at three different periods during the 24-hour day were also taken into account in arriving at the most adverse load effect for each span and influence line.

14.2.3 Flowing traffic

For flowing traffic a number of different vehicle types conforming to the *C&U regulations* were considered together with two special 6-axle vehicles with gross

Table 14.1. Impact factors for flowing traffic

Span (m)	Moment impact factor		Shear impact factor	
	Single lane	2 lanes	Single lane	2 lanes
5	1.70	1.28	1.40	1.28
10	1.55	1.26	1.35	1.26
20	1.40	1.22	1.25	1.22
30	1.40	1.18	1.20	1.18
40	1.40	1.14	1.20	1.14
50	1.40	1.10	1.20	1.10
75	1.40	1.10	1.20	1.10
100	1.40	1.10	1.20	1.10
150	1.40	1.10	1.20	1.10
200	1.40	1.10	1.20	1.10

weights of 46 tonnes and 44 tonnes. All the vehicles were run individually over the influence lines for spans up to 50 m, and in tandem pairs with a gap of 5 m between them. The effects of two vehicles side by side were also investigated. The weight distributions of the vehicles were derived from data for the nearest similar type recorded in the 14-day Warrington data. Each vehicle was assumed to have a twin-mode weight distribution representing the laden and unladen state; allowances were also made to take account of dynamic effects. From the statistical parameters of the distributions of the load effects it was possible to derive target values for the single, tandem and side-by-side vehicle effects for different spans and influence lines and for the worst of the three periods of traffic flow, and to derive these for the different return periods. The latter were based on assumed annual vehicle-flow rates with 1.0 per cent of fast moving lorries being closely spaced in convoy and 0.1 per cent of them overtaking and being closely spaced side by side. The overall effects for each span were obtained by selecting the maximum effect from the three different traffic flow conditions during the 24-hour day. Dynamic effects were applied to the flowing traffic for both single and 2-lane loading and for both moment and shear cases, with the moment and shear factors being the same for 2-lane loading. The impact factors were generally constant beyond about 40 m span as shown in Table 14.1; they were based on the work described in Chapter 13 on the development of the traffic loading models for the *Eurocode*.

14.2.4 Adjustment factors—Load Model 1

The adjustment factors α for the tandem axle and the UDL systems were derived by trial and error from a comparison of the total ENV 1991-3 [1] load effects, calculated using proposed adjustment factors, with target values derived from the simulations described above. These target values were based on the more onerous of the effects of either congested or flowing traffic for each influence line. The adjustment factors were progressively modified by an optimisation process to minimise the difference between the 'target' effects and the 'load model' effects. The adjustment factor for the tandem-axle load was biased towards matching the effects for the shorter

Table 14.2. Initial adjustment factors for LM1 [2]

	Tandem system	UDL system
Lane 1	0.85	0.70
Lane 2	0.85	0.70
Lanes 3 and 4		2.40
Remaining area		0.70

spans for a single lane, whereas the adjustment factor for the distributed load system was biased towards matching the longer spans with two lanes of traffic. A comparison of the target and load model effects using the final selected adjustment factors showed that the match was good for single-lane traffic for maximum mid-span bending moments. For 2-lane traffic the match was not so good, especially for the shorter spans.

Adjustment factors were also determined for lanes 3 and 4 on the basis that there should be the least alteration in factored design load requirements between the NAD and the current UK requirements in BD 37/88 [3]. The loading in the latter was based upon the average expected load effects from traffic containing a high proportion of HGVs, whereas the traffic models used to derive the implicit lane variations given in ENV 1991-3 contained significantly lower proportions of HGVs. The proposed adjustment factors resulted in the distributed loading in lanes 3 and 4 exceeding that in lane 2. This was explained by the fact that on a 2-lane carriageway the loading in lane 2 will be considerably diluted by light vehicles, whereas on a 3-lane carriageway only one of the lanes can be deemed to be carrying diluted traffic and hence the other two will still both be more heavily loaded. The actual designation of the lane numbering is immaterial since the designer has to place each lane load in its most onerous position. The adjustment factors proposed initially for Load Model 1 are shown in Table 14.2.

14.2.5 Adjustment factors—Load Model 2

An examination of the distribution of the axle weights recorded on the worst lane on the M6 at Warrington showed that the upper tail was well matched by a normal distribution. From this the 1000-year characteristic axle load, including the 1.7 dynamic factor from the derivation of the ENV 1991-3 loading, was found to be 320 kN, or an equivalent adjustment factor of 0.85 on the axle weight value given in the ENV for lane 1. The proposal was that the same axle load should apply to all lanes.

14.2.6 Reduction factors

The reduction factors ψ, which also cover the factors to be applied to loads in combination with other loads, are applied to characteristic load values to obtain values which are appropriate to apply in cases where infrequent or frequently occurring events are being considered. The reduction factors for inclusion in the NAD were calculated by obtaining the statistically rare and frequent effects for a combination of influence lines and span ranges and comparing them with the corresponding characteristic effects. The values were obtained by repeating the calculations for the characteristic loads described above but using the appropriate loads for the rare and frequent events. For flowing traffic, and based on a normal distribution, the number of standard deviations to reach the design value were reduced in accordance with the number of vehicles passing in the reduced time period. For the congested traffic case the return periods in the extremal distributions were similarly reduced. The process of obtaining the optimum values of the reduction factors was similar to that used in the determination of the adjustment factors, where the 'target' values of the load effects derived from the M6 traffic data were compared with the 'model' values derived from the traffic models in ENV 1991-3 [1]. It was found that the rare effects for lanes 1 and 2 were 98 per cent and 70 per cent of the characteristic effects for the tandem-axle system and the distributed load system respectively. The frequent effects were 94 per cent and 60 per cent of the characteristic effects for the tandem-axle and distributed loads respectively. The reduction factors for the single-axle load (LM2) were found to be 87 per cent and 75 per cent of the characteristic values for rare and frequent events respectively.

14.2.7 Abnormal vehicles—Load Model 3

In addition to the work on normal loading, Flint and Neill also looked at the models for abnormal vehicles. The aim was to find out which of the SVs from Load Model 3 with the associated frequent loading from Load Model 1 together with the ULS partial factor most closely represented the HB vehicle together with its associated HA distributed load and ULS partial factor. In addition, the effects of the NAD version of Load Model 1 were compared with the effects of the HB vehicle and with the Load Model 3 SVs. Five 45-unit HB vehicles were included in the comparisons with the inner axle spacing ranging from 6 to 26 m, together with 17 SVs selected from the list given in ENV 1991-3. All the vehicles with their associated distributed loading were run on a traffic simulation program to obtain the maximum static mid-span bending moments on a simply-supported span for a single lane of traffic and for spans from 5 to 200 m. The support moments generated at the central support of a two-span continuous beam were also studied, but the results were only valid for a total span length of 50 m plus the vehicle length as the associated distributed load before and after the vehicle was omitted. The most adverse of the HB vehicle effects were selected and the results from all the different vehicles were normalised with respect to the HB plus associated HA effects.

It was shown that the NAD Load Model 1 and the BS 5400 HA load effects were less than the effects from the HB vehicles plus associated HA for all spans, and that the effects from HA loading were less than the NAD LM1 effects for all spans. For the *Eurocode* SVs it was shown that a single vehicle on its own governed for spans up to 50 m where the effect of the associated distributed loading started to become significant. At spans greater than 75 m the effect of the associated distributed loading started to become the dominant factor. The SV most resembling the 45-unit HB vehicle was found to be the 1500/200 8-axle vehicle with an overall weight of 1500 kN and length of 10.5 m; the corresponding HB vehicle had an inner axle spacing of 6 m with an overall length of 9.6 m and an overall weight of 1800 kN. The effects of the SV loading became less than the effects of HB loading at longer spans where the contribution of the frequent value of the associated LM1 distributed loading was less than the HA component. It was found that the match was not so good when considering the central support influence line where the effects of the shorter length 1500 kN SVs were considerably lower than the maximum effects from the 45-unit HB vehicle with its adjustable length. It was concluded that it was only possible to properly match the HB vehicle with a SV on simply supported bridges.

As a result of their initial studies Flint and Neill recommended that the following SVs should be used to replace the HB vehicle in the NAD:

45-unit HB—SV 1500/200
37.5-unit HB—SV 1200/150
30-unit HB—SV 900/150

In the above the first figure of the SV description represents the weight of the vehicle in kN and the second figure the weight of the heaviest axle in kN. It was noted that the vehicles selected were relatively short and did not adequately represent low-loaders or road trains. It was suggested that if it was considered important to cover such vehicles then the SV 2400/200 would provide the best model. It was also noted that checks against 30-unit HB were not necessary for designs conforming to the NAD requirements.

14.3 Final parametric calibrations

14.3.1 Scope

The calibration of the ENV against BD 37/88 covered span lengths from 3 to 200 m and involved the use of the same influence lines to calculate load effects as were used in the derivation of the ENV loading criteria (see Chapter 13). Because of the substantial differences between the ENV method of determining the numbers of notional traffic lanes using a fixed 3 m wide lane and a remaining area, and the UK practice where notional lanes can vary from 2.5 to 3.7 m with no remaining area, it was felt to be necessary to consider a much wider range of carriageway and notional lane widths than used in the initial studies. Thus 2- to 6-lane carriageways were considered together with single-lane carriageways, and standard design lane widths as well as a spectrum of non-standard lane widths were used. The calibration took account of the γ_{f3} factor of 1.1 which is included in the material design parts of British Standards but not in the equivalent *Eurocodes*. Because BD 37/88 contains a 10 per cent contingency allowance for future increases in traffic loads it was decided that the partial traffic load factors in the ENV should be increased by 10 per cent to reflect this and they thus became 1.5 and 1.1 for the ultimate and serviceability limit states respectively.

14.3.2 Adjustment factors (α)

By comparing the bending moment and shear effects for the different spans and influence lines derived from the application of the ENV and BD 37/88 loadings, the values of the adjustment factors which give the best fits for both the distributed and tandem axle loadings for LM1 were obtained. However, it was not possible to

Table 14.3. Final adjustment factors for LM1 [4]

	Tandem system	UDL system
Lane 1	0.844	0.40
Lane 2	1.267	1.44
Lane 3	1.267	1.44
Other lanes		1.44
Remaining area		1.44

obtain exact fits for all span lengths and carriageway widths and so the α values were optimised to give the best overall correlation. In doing this higher weighting was given to the wider bridges at the longer spans and the narrower bridges at the shorter spans. The resulting set of ratios of NAD to BD 37/88 effects incorporating the proposed adjustment factors was generally in the range 0.9 to 1.1. The values of the adjustment factors were progressively modified with some loss of accuracy in the calibrations in order to simplify the application of the factors and the loading in the different lanes. It was found that a satisfactory correlation between the NAD and the BD 37/88 load effects could be obtained by using adjustment factors which with the ENV characteristic loads gave a uniform UDL which could be applied over the whole area of a multi-lane carriageway. Details of the final adjustment factors derived for LM1 are shown in Table 14.3.

Because of the decision to calibrate against BD 37/88 it was also necessary to revise the value of the single-axle load model LM2 which is modified by the lane 1 tandem-axle adjustment factor. In addition it was found to be necessary to use the single-axle model to fill a shortfall in the span range from about 3 to 7 m where the tandem axle in LM1 failed to reproduce the effects of the BD 37/88 HA KEL. The single-axle load also had to reproduce, as far as possible, the BD 37/88 HB axle and wheel loads. To meet these requirements, it was calculated that the resulting single-axle load should be 380 kN, which meant that with the application of the lane 1 adjustment factor of 0.844 the boxed value of the single-axle load should be 450 kN.

14.3.3 Reduction factors (ψ)

Although the statistical procedures used in the initial studies, which were based on measurements of actual traffic, were not used to derive the final adjustment factors they were, nevertheless, used to determine the ψ_1' and ψ_1 factors for determining the loads for infrequent and frequent loading situations. The derivation of the values of the reduction factors for both the LM1 and LM2 load models are as described earlier for the initial study.

14.3.4 Abnormal vehicles—Load Model 3

It was shown in the initial calibration study that none of the longer single-load models in the garage of SVs in ENV 1991-3 produced load effects which had a consistent relationship with those due to the HB vehicle in BD 37/88. To improve the correlation by matching the variable wheelbase of the HB vehicle it was decided to use a pair of SVs with a certain amount of flexibility in the spacing between them.

It was found that this approach helped to solve virtually all the calibration problems which had been encountered when using just a single vehicle. The pairs of vehicles, with 150 kN axles, were selected to have the same overall weights as the weights of the HB vehicles which were applicable to the different classes of road in the UK. Thus for motorways and trunk roads designed for 45 units of HB loading, two no. 900 kN SVs were to be applied; for principal roads designed for 37.5 HB units a 900 kN and a 600 kN SV were to be applied. For other public roads which are designed for 30 units of HB loading two no. 600 kN SVs were required. The minimum gap between each pair of vehicles was 1.5 m with a 25 m gap between the vehicles and the associated LM1 loading. However, it was found that the reduction factors to be applied to the associated loading when combined with the LM3 loading needed to be higher than the 'frequent' values as prescribed in the ENV in order to match the BD 37/88 HB and associated HA loading. Thus the reduction factors on the coincidental LM1 loading were increased to 1.0 for the tandem-axle system loading and 0.88 for the distributed load component.

14.3.5 Secondary loads—braking

A comparison was made of the braking and acceleration forces in the ENV and BD 37/88 respectively. This showed that at the ULS for loaded lengths up to 200 m the ENV forces were between 0.6 and 1.15 times those from BD 37/88. For a single-lane carriageway the corresponding ratio was between 0.5 and 0.85. The ENV values have been retained but a minimum value of 300 kN has been added; this corresponds to the BD 37/88 skidding load.

14.3.6 Secondary loads—centrifugal forces

A comparison of the ULS factored centrifugal forces derived from the ENV and BD 37/88 was done for different numbers of lanes and various horizontal radii. The ratios varied considerably but generally lay in the range 0.3 to 0.9. The centrifugal force requirements in the ENV have been retained as they were, but there is a note in the NAD which draws attention to the fact that they can be significantly less onerous than those in the DMRB.

14.3.7 Accidental loads—collision forces on piers

The unfactored forces of 1000 kN in the direction of travel and 500 kN at right angles to the direction of travel are the same in both the ENV and the current UK requirements in BD 60/94 [5]. However, the partial factor in the ENV on accidental loads is taken as 1.0 whereas using the BD 37/88 factors of 1.5 × 1.1 (i.e. 1.65) the factored loads come to 1650 kN and 825 kN respectively. These are the values which have been inserted in the NAD, but there is also a general requirement to comply with the criteria in the DMRB which involves the application of residual loads as well.

14.3.8 Accidental loads—vehicles on footways

Even allowing for the revised axle load factors in the NAD, the resulting ENV factored load for a tandem axle at ULS is slightly less onerous than the heaviest axle load from the BD 37/88 accidental axle load model. However, the NAD does not modify the ENV requirements apart from the increase in the adjustment factors that has been mentioned before.

14.4 Published version of the UK NAD

14.4.1 Introduction

The background to the development of the NAD and the derivation of the boxed values have been described above. In total the NAD was developed from the following activities:

(i) a textual examination of ENV 1991-3 [1];
(ii) parametric calibrations of ENV 1991-3 [1] against current UK bridge loading requirements, as specified in BD 37/88 [3], and against data from a survey of current traffic;
(iii) trial calculations.

In addition to the boxed values, the NAD makes reference to certain national documents which are needed to support the ENV. In particular it makes reference to the *DMRB* [6] as published by the Highways Agency. A table is included in the NAD which brings together all the boxed values which affect the various factors and values of the characteristic loads themselves. The table shows both the original ENV values and the replacement NAD values. Below is a summary of the more important changes that have been made to ensure that loading derived from the ENV, as modified by the NAD, is in good agreement with current UK highway bridge loading.

14.4.2 Partial safety factors (γ)

The partial factors on traffic loading have been increased from 1.35 and 1.0 to 1.5 and 1.1 for the ULS and SLS respectively to allow for the γ_{f3} factor of 1.1 which is applied when using the British Standard materials codes. In the case of accidental actions only the ULS is to be considered and the partial factor has been kept at 1.0. Where traffic loading is to be combined with other forms of loading the combination factors ψ_0 are normally based on the frequent ψ values for both the ULS and the SLS. However, in the case of the LM3 loading the coincident LM1 loading is to consist of 1.0 times the tandem axle loads and 0.88 times the distributed load. Load Models 2 and 4 are considered on their own and so do not require combination factors.

14.4.3 Adjustment factors (α)

The main change from the ENV, where the adjustment factors are usually taken as unity, is the inclusion of a set of factors for the different lanes in a multi-lane carriageway. These have been the main vehicle for matching the ENV loading to the UK loading. The resulting changes allow the retention of the ENV concept of 3 m wide notional lanes and the LM1 model composed of the tandem-axle system together with a distributed load. However, in carrying out the calibration exercise it was found possible to simplify the distributed loading requirements so that the factored UDL is a constant value which is applied over the whole of the carriageway

Table 14.4. Factors on characteristic actions for combination of actions for highway bridges (ULS) [1]

Load type and system	Groups of loads									
	Group 1		Group 2		Group 3		Group 4		Group 5	
a. Permanent actions										
Direct actions										
self-weight: concrete	1.35		1.35		1.35		1.35		1.35	
self-weight: steel	1.20		1.20		1.20		1.20		1.20	
superload: surfacing	1.35		1.35		1.35		1.35		1.35	
superload: other	1.35		1.35		1.35		1.35		1.35	
ground conditions	1.20		1.20		1.20		1.20		1.20	
Indirect actions										
prestress/creep	1.00		1.00		1.00		1.00		1.00	
settlement	1.00 (0)		1.00 (0)		1.00 (0)		1.00 (0)		1.00 (0)	
b. Variable actions										
Carriageway	TS	UDL	TS	UDL					TS	UDL
Vertical forces										
main loading system: LM1										
lane 1	**1.27**	**0.60**	1.19	0.36	0		0		1.27	0.53
lane 2	**1.90**	**2.16**	1.79	1.30	0		0		1.90	1.90
lane 3	**1.90**	**2.16**	1.79	1.30	0		0		1.90	1.90
other lanes	**0**	**2.16**	0	1.30	0		0		0	1.90
remaining area	**0**	**2.16**	0	1.30	0		0		0	1.90
special vehicles: LM3	0		0		0		0		**1.50**	
single axle: LM2	$1.50\beta_Q$		$1.50\beta_Q$		$1.50\beta_Q$		$1.50\beta_Q$		$1.50\beta_Q$	
crowd loading: LM4	0		0		0		**1.50**		0	
Horizontal forces										
braking and acceleration	0		**1.50**		0		0		0	
centrifugal forces	0		**1.50**		0		0		0	
Footway and cycle tracks	UDL	S veh.			UDL	S veh.	UDL	S veh.		
	0.75	0	0	0	**1.50**	**0**	**1.50**	1.50	0	0
Wind										
F_{Wk} or F_{Wn}	0.45		0		0		0		0	
or F_W	1.50		0		0		0		0	
Temperature T	0		0		0		0		0	
c. Accidental actions										
Under bridge	0		1.00		1.00		0		1.00	
On bridge	0		0		0		1.00		0	

Notes: S veh. = service vehicle.

area and is also constant over the whole span range. Also the adjustment factors on the tandem-axle load now produce equal loading in the first two notional lanes of the carriageway.

14.4.4 Reduction factors (ψ)

The reduction factors required to convert characteristic load values to values representative of infrequent and frequent events have also been increased. For example for infrequent events the factors have been increased to 0.98 for the LM1 tandem axles, to 0.7 for the LM1 distributed load and to 0.87 for the LM2 single axle. The reduction factor for the LM3 SV is retained at 1.0.

14.4.5 Tables of factors

Besides giving separate tables for partial factors (γ), adjustment factors (α) and combination and reduction factors (ψ), the NAD also provides tables in which all these various factors have been combined. An example is given in Table 14.4 which shows the combined factors ($\alpha\gamma\psi$) to be applied to the relevant characteristic loads for different combinations of actions at the ULS. Other tables cover the SLS for rare (characteristic), infrequent and frequent load combinations. The bold type in the tables indicates the dominant component action. Thus under the 'Groups of loads' heading, Group 1 is concerned with traffic loading together with wind and temperature loading. Group 2 is concerned with horizontal loads such as braking and centrifugal force. Group 3 covers footway and cycle-track loading and Group 4 crowd loading. Abnormal loading, involving the application of the SVs, is covered by Group 5. Footnotes (which are not shown here) define how the various loads are to be applied; thus wind can be applied with traffic loading or it can be applied on its own with the factors being specified for each case. These tables will help to ensure that the appropriate factors with their correct values are applied for any required group of loads when in particular combinations.

References

1. British Standards Institution (2000) DD ENV 1991-3: 2000. *Eurocode 1: Basis of design and actions on structures—Part 3: Traffic loads on bridges*. London: BSI.
2. Flint and Neill Partnership (1995) *Eurocode 1—Part 3: Traffic loads on bridges. Development of the quantitative provisions in the National Application Document*. Unpublished report for the Highways Agency.
3. Department of Transport (1989) Departmental standard BD 37/88, *Loads for highway bridges*. London: DTp.
4. Flint and Neill Partnership (2000) ENV 1991-3 (*Eurocode* 1 Part 3): *Traffic loads on bridges. Draft background report on the development of the quantitative provisions in the National Application Document*. Unpublished report for the Highways Agency.
5. Highways Agency (1994) *DMRB*, Vol. 1, Section 3, Part 5, BD 60/94, *The design of highway bridges for vehicle collision loads*. London: HMSO.
6. Highways Agency *Design manual for roads and bridges*. Norwich: HMSO.

Appendix
Use of OPTAX program

A1 Introduction

The computer program OPTAX had the facility to determine the optimum axle weights and spacings of a proposed vehicle traversing a simply supported span, such that the maximum bending moments and shear forces caused by various trains of vehicles, including the proposed vehicle, did not exceed those due to a control vehicle for any span. However, for the work involved with developing loading models the use of OPTAX was limited to finding the envelope of maximum bending moments and shear forces due to various trains of vehicles, including a particular existing or proposed vehicle, hereinafter referred to as the 'test vehicle', over a range of simply supported spans.

A2 Vehicle trains

Four distinct trains of vehicles were used as the loading cases for each span, and the program selected and identified the loading train producing the worst bending moment and shear force for the span in question. Details of the four loading trains are as follows:

(i) test vehicle on its own;
(ii) test vehicle preceded and followed by a 24.4-tonne standard vehicle with the dimensions shown in Fig. A1 [1];
(iii) three test vehicles in a row;
(iv) eight test vehicles in a row.

The term 'test vehicle' covered any vehicle which was allowed to operate under the existing or proposed *C&U regulations*, including possible European vehicles. Thus to find the envelope of maximum loading effects it was necessary to repeat the process for a number of vehicles which had been identified as the worst of their class. The short trains were pertinent for the shorter spans, whereas the longer trains were

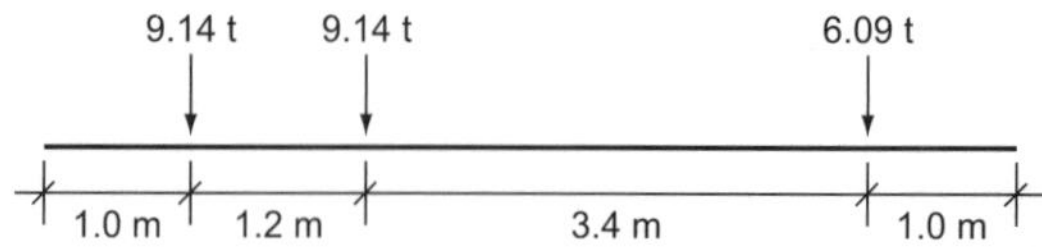

Figure A1. Axle arrangement of 24.4-tonne vehicle [1]

more critical for the longer spans. The three-axle standard vehicle was chosen as being a critical example of its type as far as the loading effects were concerned.

A3 Vehicle spacing

It was assumed that there was a 1 m gap between each vehicle in a train and that each vehicle had an overhang of 1 m at the front and rear. Thus there was a 3 m spacing between the rear axle of one vehicle and the front axle of the following vehicle. These overhangs and spacings were believed to be typical of the vehicles at the time and typical of normal jam situations.

A4 Impact factors

An impact factor of 1.8 was used and included in the OPTAX calculations. This was only applied to one axle and only in the loading case with a single test vehicle. The program automatically applied the impact factor to the axle producing the worst effect as far as bending moments and shears were concerned.

Reference

1. BES Division (1983) *Revision of short span loading*, DTp paper (presented at Seminar on revision of highway loading).

Index

Page numbers in italics refer to tables or figures.

abnormal vehicles
  see also HB vehicle loadings; HB vehicles
  increase in numbers 21
  loadings
    modelling 101–102, 104–105, 108–113, *111*
    spans 102
  long-spans, loadings 102–104
  special order vehicles 100, 101
    effects of *10*
    loading models 106, *107*, 108, *108*
    movement regulations 101
  special types general order vehicles
    catagories 100
    configurations 114
    contingency margins 104
    dynamic load factors 112, *113*
    effects of *10*
    loading models 104–106, 113–114
    long-span loadings 102–104
    movement regulations 101
    spacing 105
  special vehicles
    axle lift-off factor 113
    configurations 110, *111*
    dynamic amplification factors 111–112, *113*
    Eurocode 4
    HB conversion charts 115
    loading assumptions 114
    loading models 109–110, *110–111*
    overload factors 110–111
    partial load factors 113
    reserve factors 114
  Type C, standards 18–19, *19*, 99
accidental wheel loadings 29, *72*, 73
accidents
  see also collisions
  causing jams 59
articulated vehicles
  axle spacing *129*
  axle weights, 1904–1999 *128*
  gross weights, 1998 *129*
assessment codes, safe load capacity 75–76
assessment of highway bridges and structures, The (DTp) *9*
assessment live loading
  C&U regulations 91, *92*
  definitions 96
  dynamic load models
    amplification factors 78–79
    data 78
  existing bridges 2–3
  lateral bunching, dynamic effect 80
  load reduction factors 80–81, *81*
  methodology 76–77, *76*
  probabilistic load models 79
  reduction factors 93–94, *94*
  reliability analyses 79–80
  static load models
    traffic modelling 77–78
    vehicle data 77
assessment loadings, vehicle trains 84–86, *84*, *85*
Authorisation of special types – general order, limits 21

axles
lift-off factor 113
single loadings 89, 91–92
weights, 1904–1999 *128*

bending moments (BM), OPTAX program 46–47, 126
bogies, trailers, spacing 26
box girder bridges, steel *9*, 30, 42
braking loads
see also skidding loads
HA vehicles 26, 38–40, *39*
HB vehicles 40
introduction 5, 19, 26
National Application Documents 154
Traffic loads on bridges ENV 1991-3 144, *144*
values 73
bridge loadings, controls, pro-active 10
bridges, future loadings 7, 10
British Engineering Standards Association (*later* British Standards Institute), steel bridges 6, 15, 17
British Rail, Operation bridgeguard 86
British Standards Institute (BSI) *see* BS 153; BS 5400; National Application Documents
British Waterways, Operation bridgeguard 86
BS 153
Part 3
1923 *8*, 15, 17
loading trains *8*, 17–18, *17*, *18*
Type A loadings *18*
Type B loadings 17, *18*
Type C loadings 18–19, *19*
Part 3A 1954 21–27, *22*, *25*
adoption 6, *8*
revisions *8–9*
Part 3A 1972, metrication 28
BS 5400: Part 2: 1978
amendments 42–43
collision loads 118
HA loading curves 34–35, *35*
issue 6, *9*, 33, 56, 74
limit state design 33–34
long-span bridges, revised curve 42–43, *43*
BS 6779: Part 1: 1998, Appendix A 119

C&U see Construction & Use Regulations
carriageways
lanes
loadings 24–25
notional 89, 142–143
standard widths 50, 71, 73
vehicle distribution 66
width factors 89, *90*, 93
number of 5
occupancy, HB loadings 36
remaining areas 142
roughness values, dynamic factors 139–140, *140*
cars, loading allowances 88, *88*
cast iron bridges, assessment 83
centrifugal forces
curved bridges 29, 37–38, *38*, 73, 89–90
National Application Documents 154
significance 5
Traffic loads on bridges ENV 1991-3 144–145
characteristic values, load effects 34, 56, *57*
collision loadings
computer simulations 121–122
decks 120, *120*
guard rails 41, *41*
main load components 120–121
parapets 29, 40–41, 42, *72*, 73, 119
safety fences 41, 117–118
supports *72*, 73, 118–119, *119*
prevention 29, 41, 42
combination loadings
HA vehicles 103, *103*, *104*
factors *72*, 73
computer analysis
OPTAX program 46–47, 126, 159–160
structures 5
computer simulations, collision loadings 121–122
concrete
prestressed beams, computer simulations 121–122
serviceability limit states 34
slabs, and loading specifications 24, *25*
Concrete Institute Report 1918 *8*
highway classifications 11–12
Construction & Use Regulations (*C&U*)
1973 126–127, *127*
assessment live loadings 91, *92*
axle limits 56, 85
introduction 126
vehicle design 3
weight limits 21, 56, 85, 86
containers, 40 ft ISO 51
contingency margins, EC vehicle proposals 52, 91
contra-flow working 5
convoys
see also traffic jams
generation of test 56
HA loadings 48, *49*
cumulative distribution functions (CDF), traffic modelling 77
curved bridges, centrifugal forces 29, 37–38, *38*, 73, 89–90

DAF *see* dynamic amplification factors
data collection, principles 7, 10
dead loadings 3
decks, collision loadings 120, *120*, 143
Department of the Environment (DoE)
  see also Department of Transport; Highways Agency; Ministry of Transport
  Standard highway loadings BE 5/73 *9*, 28–29, 86–87
Department of Transport (DTp)
  see also Department of the Environment; Highways Agency; Ministry of Transport
  The assessment of highway bridges and structures 9
  bridge rehabilitation programme 75–76
  Design manual for roads and bridges BD 21/93 93
  Design manual for roads and bridges BD 48/93 120–121
  Design manual for roads and bridges BD 60/94 121
  Design manual for roads and bridges BA 16/97 95–96
  Design manual for roads and bridges BD 21/97 93–95, *94*
  Loads for highway bridges BD 14 *9*
  Loads for highway bridges BD 37/88, NAD cross-calibration 152–153, *153*
  Standard BD 14/82 *9*, 41–42, 118
  Standard BD 21/84 45–46, 87–90, *88*
    Advice note BA 16/84 90
    Amendment No.1 90–93, *92*
  Standard BD 37/88 67
    composite version 69
  Standard highway loadings 9
  Technical memorandum BE1/77, 1977 30
design, traffic modelling 87
Design manual for roads and bridges BD 21/93 (DTp) 93
Design manual for roads and bridges BD 48/93 (DTp) 120–121
Design manual for roads and bridges BD 60/94 (DTp) 121
Design manual for roads and bridges BA 16/97 (DTp) 95–96
Design manual for roads and bridges BD 21/97 (DTp) 93–95, *94*
Design manual for roads and bridges BD 21/01 (HA) 96
Design manual for roads and bridges BD 37/01 (HA) 74
Design manual for roads and bridges BD 86/01 (HA) 113–115
DoE *see* Department of the Environment
DTp *see* Department of Transport
dynamic amplification factors (DAF) 78–79
  special vehicles 111–112, *113*

EC *see* European Commission Directives
ENV *see* EuroNorm Vornorms
equivalent distributed loads, 1000-year return extrapolations 139, *139*
equivalent loading curves, MoWT Memorandum 20
equivalent loading curves *16*
  HA loadings 22–23, *22*, *24*
    short-span members 23
  introduction 13, 15
  long-span bridges 42–43, *43*
  span limitations 15
equivalent loading models, construction 21–22
equivalent uniformly distributed loads (EUDL) 3
  determination
    extremal distribution 63, *64*
    lane load factors 64–65, *65*
    method 62
    normal distribution 63, *64*
  modelling effectiveness 4
EUDL *see* equivalent uniformly distributed loads
Eurocodes
  see also EuroNorm Vornorms; National Application Documents
  1: Actions of structures 133
    Part 3: Traffic loads on bridges 9
  purpose 133
  tandem axle arrangements 4
  UK contributions to 7
EuroNorm Vornorms (ENV) 133
  see also National Application Documents
  1991-3 *Traffic loads on bridges*
    application of loading 142–143
    braking loadings 144, *144*
    centrifugal forces 144–145
    collision loadings 143–144
    concentrated load models 134, *135*, 137–142, *139*, *140*, *141*
    crowd loading 137
    development 134
    single-axle load model 134–135, *136*
    special vehicles 135–136, *136*, *137*
European Commission Directives, 85/3/EEC, vehicle weights 51–52, *52*, 91, 95–96, 128, *129*
existing bridges
  assessment loadings 2–3
  loading 2–3
extreme loadings
  computer simulations 46–47
  factors 46

fatigue, limit state design 34
fire engines, loading allowances 88, *88*
footbridges, supports, impact protection 30
footways
  wheel loadings 27, 143–144, 154–155
    accidental 29

gantries, supports, impact protection 30
guard rails, collision loadings 41, *41*

HA vehicle loadings
  see also HA vehicles
  combinations, factors *72*, 73
  concrete slabs 24, *25*
  equivalent loading curves 22–23, *22*, *24*
    adjustments to 29
    multi-lane carriageways 34–35, *35*
  equivalent uniformly distributed load
    design values 66–67
    determination 62–65
  increased 27
  introduced *8*
  lateral bunching 48, 50, 95
  live loads 89
    partial factors 36–37
  long-spans
    extremal distribution effects 63, *64*
    lane factors 70–71, *71*, 72
    normal distribution effects 63, *64*
    traffic jams 56, 59–60
    traffic spacing 55
    vehicle characteristics 57–59, *58*
  short-spans
    equivalent loading curves 35
    lane factors 70–71, *71*, 72
    lateral bunching 48, 50
    specific 45
    ultimate design 46
    uniformly distributed 50–51, *51*
  uniformly distributed, curves 50–51, *51*, 69–70, *70*
  vehicle characteristics 57–58, *58*
HA vehicles
  braking forces 26, 38–40, *39*
  convoys 48, *49*, 56
  impact factors 35
  overloading factors 47–48, *48*
    axles 48, *49*
  traffic streams, assembly of typical 61, *62*
HB vehicle loadings
  see also abnormal loads; HB vehicles; special vehicles
  arrangement *28*
  combinations, factors *72*, 73
  introduced *8*, 26
  live loads 89
    partial factors 36–37
  modelling 108–109
  motorway bridges 29
  overstress allowances 27
  wheels 36
HB vehicles
  see also abnormal vehicles
  braking forces 40
  definitions 99–100
    proposed *106*
  designated roads 27
  impact factors 26
  inter-axle distances 35
  lane occupancy 36, 70
  pneumatic tyres 29
  public roads 42
  special vehicles, conversion charts 115
  trailers
    bogies 26
    weights 26
  trunk road bridges 29
heavy goods vehicles (HGV)
  see also HA vehicles; HB vehicles
  definitions 131
  flow variations 65–66
  spot weight checks 131
Heavy Motor Car Order, 1904 126
highway classifications, Concrete Institute Report 11–12
Highways Agency
  see also Department of the Environment; Department of Transport; Ministry of Transport
  Design manual for roads and bridges BD 21/01 96
  Design manual for roads and bridges BD 37/01 74
  Design manual for roads and bridges BD 86/01 113–115
  responsibilities 6–7
  Standard BD 37/01 74

impact factors
  HA loadings 35
    multiple vehicles 47
  HB loadings 26
  span, increasing 18
  Standard Loading Trains 13, 15
Institution of Civil Engineers
  Code of Practice 6, 18–19
  steel bridges, Type C Abnormal Loads 18–19, *19*
Institution of Structural Engineers, steel bridges, standards 6, 18–19

jams *see* traffic jams
Joint Committee on Loads on Highway Bridges, 1918 Report 125

knife-edge loads (KEL) 3
and equivalent loading models 21–22

lanes
live load factors 64–65, *65*, 66–67, *67*, 95
occupancy, HB loadings 36, 70
standard widths 50, 71, 73
vehicle distribution 66
width factors 89, *90*, 93
lateral bunching
assessment live loading, dynamic effect 80
HA vehicle loadings 48, 50, 95
limit state design
adoption 5, 87
characteristic values 34
fatigue 34
principles 33–34
serviceability limit state 33–34
ultimate limit state 33–34
live loads 2–3
lane factors 64–65, *65*, 66–67, *67*, 95
partial factors 143
HA loadings 36–37
HB loadings 36–37
secondary 73
load assessments, traffic modelling 87–88, *88*
load combinations
definitions 37
partial factors *38*, *72*, 73
recognition 30
reduction factors 143
load effects, characteristic values 34, 56, *57*
load reduction factors 87–88, *88*
assessment live loading 80–81, *81*
loading models
equivalent uniformly distributed loads 3, 4
first 6
HA loading basis 27
HB arrangement *28*
knife-edge loads 3, 4
special order vehicles 106, *107*, 108
amplification factors *108*
special types general order vehicles 104–106
special vehicles, development 109–110, *110*
statistical approach 5
tandem axle arrangements 4
loading trains *see* Standard Loading Trains
Loads for highway bridges BD 14 (DTp) *9*
Loads for highway bridges BD 37/88, NAD cross-calibration 152–153, *153*
local authorities, loading standards 5–6, *8*, 11
Locomotives Act, 1861 126
Locomotives Act, 1898 126
long-spans
abnormal vehicles, loadings 102–104
HA loadings
lane factors 70–71, *71*, 72
traffic jams 56, 59–60
traffic spacing 55
vehicle characteristics 57–59, *58*

masonary arch bridges
axle lift-off 113
MEXE assessment 83
Merrison Report 1973 30, 42, 55
MEXE assessment, masonary arch bridges 83
Ministry of Transport (MoT)
see also Department of the Environment; Department of Transport; Highways Agency
BSI standards adopted 6
Equivalent Loading Curve *16*
introduction 13, 15
span limitations 15
establishment 2, *8*
Interim memo IM 10, 1970 27–28
national loading model, first 6
Report on administration of the Road Fund for year 1921–1922 8
Standard highway loadings Memo 771 *8–9*, 27
Standard of loading for highway bridges 8
Technical memorandum (bridges) BE 4 83–86
Ministry of War Transport (MoWT), Memorandum No 577 *8*, 20
monitoring traffic loading, WIM devices 130
Motor vehicles (construction and use) regulations, introduction 126
motorway bridges, HB loadings 29
movement regulations
special order vehicles 101
special types general order vehicles 101, 115
MoWT *see* Ministry of War Transport

National Application Documents (NAD) 133–134, 147
calibration studies 147–148
adjustment factors (LM1) 149–150, *150*
adjustment factors (LM2) 150
congested traffic 148
flowing traffic 148–149, *149*
influence lines considered 148
reduction factors 150–151
special vehicles (LM3) 151–152
calibrations
adjustment factors (LM1) 152–153, *153*
adjustment factors (LM2) 153

National Application Documents (NAD) (*continued*)
adjustment factors (LM3) 153–154
braking loads 154
centrifugal forces 154
support collisions 154
vehicles on footways 154–155
cross-calibration, BD 37/88 152–153, *153*
published
adjustment factors 155–156, *156*
development 155
partial safety factors 155, *156*
reduction factors 156–157, *156*

Operation bridgeguard 86
OPTAX program
bending moments 46–47, 126
shear forces 46–47, 126
use of 159–160
overload factors (OF)
HA vehicles 47–48, *48*
axles 48, *49*
special vehicles 110–111
overstress allowances, HB loadings 27

parapets, collision loadings 29, 40–41, 42, *72*, 73, 119, 143–144
partial factors
live loads 36–37, 143
load combinations *38*, *72*, 73
special vehicles 113
partial safety factors, National Application Documents 155
permanent loads 37
permissible stress design, demise 5
poor surfaces, heavy traffic, reduction factors 93–94, *94*
prestressed concrete beams, collisions, computer simulations 121–122
probabilistic load models, assessment live loading 79
public roads, HB loadings 42
public service vehicles, loading allowances 86, 88, *88*

reduction factors, assessment live loading 93–94, *94*
rehabilitation programme, 15-year 89, 118–119
remaining areas, carriageways 142
Report on administration of the Road Fund for year 1921–1922 (MoT), Standard Loading Train *8*, 13, *14*
return extrapolations, 1000-year 138–139, *139*
rigid vehicles
axle spacing *129*
axle weights, 1904–1999 *128*
gross weights, 1998 *129*
Road vehicles (authorised weight) regulations 1998 95–96, 126, 128, *129*
Road vehicles (construction and use) regulations 126, 128

safe load capacity, assessment codes 75–76
safety fences, collision loads 41, 117–118
secondary horizontal weights, significance 5
serviceability limit state (SLS), limit state design 33–34
shear forces (SF), OPTAX program 46–47, 126
short-spans
assessment live loading, methodology 76–77, *76*
HA loadings
Equivalent Loading Curves 23, *24*
lane factors 70–71, *71*, 72
lateral bunching 48, 50
specific 45
ultimate design loading 46
uniformly distributed loads 50–51, *51*
skidding loads
see also braking loads
load forces 40, *72*, 73
spacing
long-spans, HA loadings 55
Standard Loading Trains 13
vehicles, traffic streams 66
special order (SO) vehicles
see also special vehicles
definitions 100, 101
effects of *10*
loading models 106, *107*, 108
amplification factors *108*
movement regulations 101
special types general order (STGO) vehicles
see also special vehicles
categories 100
configurations 114
dynamic load factors 112, *113*
effects of *10*
loading models 104–106, 113–114
loadings, contingency margins 104
long-spans, loadings 102–104
movement regulations 101, 115
spacing 105
special vehicles (SV)
axle lift-off factor 113
classifications, ENV 1991-3 *136*
configurations 110, *111*
dynamic amplification factors 111–112, *113*
Eurocode 4
HB vehicles, conversion charts 115

loading assumptions 114
loading models 109–110, *110–111*
National Application Documents 151–152, 153–154
overload factors 110–111
partial load factors 113
reserve factors 114
Traffic loads on bridges ENV 1991-3 135–136, *136*, *137*
Standard BD 14/82 (DTp) *9*, 41–42, 118
Standard BD 21/84 (DTp) 45–46, 87–90, *88*
Advice note BA 16/84 90
Amendment No.1 90–93, *92*
Standard BD 37/88 (DTp) 67
composite version 69
Standard highway loadings BE 5/73 (DoE) *9*, 28–29, 86–87
Standard Loading Trains
see also vehicle trains
BS 153 17–18, *17*, *18*
impact factors 13, 15
introduction 2, *8*, 13
proposed, 1918 12, *12*
replacement 3–4
spacing 13
specification *14*
Type A *18*
Type B 17, *18*
Type C Abnormal Loads 18–19, *19*
standards
see also BS 153; BS 5400; BS 6779; Department of Transport; *Eurocodes*; Highways Agency; Ministry of Transport
local authorities 5–6, *8*, 11
need for *8*
professional institutions 6
statistical approach, loading models 5
steam traction engines 1
steel bridges
box girder *9*, 30, 42
standards, first 6
STGO vehicles *see* special types general order vehicles
structural analysis, by computer 5
superstructures, impact resistance 30, *72*
supports
collisions
loadings *72*, 73, 118–119, *119*, 143, 154
prevention 29, 41, 42, *72*, 73
SV *see* special vehicles

tandem axle arrangements, *Eurocodes* 4
Technical memorandum (bridges) BE 4 (MoT) 83–84
assessment vehicle trains 84–86, *84*, *85*
load distribution 85–86, *85*
Operation bridgeguard 86
Technical Memorandum (bridges) BE 5/73 (DoE) 28–29, 86–87
temperature loadings 3, 37
traction engines 1
traction loads 38
traffic growth, predictions, to 1990 60–61
traffic jams
see also convoys
causes 59
duration 59–60, 66
frequency 59–60, 66
lane selection 60
long-spans, HA loadings 56
vehicle movement 63
traffic loading, monitoring, WIM devices 130
Traffic loads on bridges ENV 1991-3 *9*
braking loadings 144, *144*
centrifugal forces 144–145
collision loadings 143–144
concentrated load model (LM1) 134, *135*
1000-year return extrapolation 138–139, *139*
calibration 142
dynamic factors 139–140, *140*
influence lines 141, *141*
loading patterns 140–141
methodology 137–138
crowd loading 137
development 134
single-axle load model (LM2) 134–135, *136*, 142
special vehicles (LM3) 135–136, *136*, *137*
classification *136*
traffic modelling 77–78
cumulative distribution functions 77
design 87
load assessments 87–88, *88*
Traffic Signs Regulations and General Directions 1994 131
traffic streams, assembly of typical 61, *62*
trailers
bogies, spacing 26
weights, maximum 26
trunk road bridges, HB loadings 29
tyres
iron 12
pneumatic 21
solid, footprint 26

ultimate limit state (ULS), limit state design 33–34

uniformly distributed loads (UDL)
  1000-year return extrapolations 139, *139*
  concept 15
  and equivalent loading models 21–22
  extremal 63, *64*
  HA loadings 50–51, *51*, 69–70, *70*
  normal 63, *64*

vehicle loadings, definitions 2
vehicle trains, assessment loadings 84–86, *84*, *85*
vehicle weights
  1904–1999 *128*
  EC directives 51–52, *52*, 91, 95–96, 128, *129*
    UK contingency margins 52, 91
  legislation, historical 125–127, *128*

weight in motion (WIM) devices, monitoring
    traffic loading 130
weight restrictions
  gross weights 130–131
  single-axle 130–131
wheel loadings
  accidental 29, *72*, 73
  footways 27, 143–144, 154–155
  single 89, 91–92
wheel loads, HB loadings 36
WIM *see* weight in motion devices
wind loadings 3, 37
World War I, traffic increases 1
World War II, bridge maintenance 2–3